Learning Resource Centre

Park Road, Uxbridge, Middlesex, UB8 1NQ

Renewals: 01895 853344

Please return this item to the LRC on or before the
last date stamped below:

620

is an imprint of

PEARSON

Harlow, England • London • New York • Boston • San Francisco • Toronto
Sydney • Tokyo • Singapore • Hong Kong • Seoul • Taipei • New Delhi
Cape Town • Madrid • Mexico City • Amsterdam • Munich • Paris • Milan

Pearson Education Limited
Edinburgh Gate
Harlow
Essex CM20 2JE
England

and Associated Companies throughout the world

Visit us on the World Wide Web at:
www.pearsoned.co.uk

First published 1996 as *Communication for Engineering Students*
Second edition published 2001
Third edition published 2011

© Longman Group Limited 1996
© Pearson Education Limited 2001, 2011

ISBN: 978-0-273-72952-5

British Library Cataloguing-in-Publication Data
A catalogue record for this book is available from the British Library

Library of Congress Cataloging-in-Publication Data
Davies, John W.
   Communication skills : a guide for engineering and applied science students / John
W. Davies and Ian K. Dunn. -- 3rd ed.
      p. cm.
   Includes index.
   ISBN 978-0-273-72952-5 (pbk.)
   1. Communication in engineering.   2. Communication of technical information.
3. Communication in science.   I. Dunn, Ian K.   II. Title.
   TA158.5.D38 2011
   620.0014--dc22

                                                                        2010042389

ARP Impression 98

Typeset in 10/12pt Sabon by 35

Printed in Great Britain by Ashford Colour Press ltd

Contents

List of checklists

Acknowledgements

We are very grateful to the following people for their advice and help:

At Coventry University: Tim Davis and Geoff Powell (Civil Engineering), Malcolm Blake (Engineering), Richard Newman (Careers Adviser).

At Aberystwyth University: Ann Robertson (Computer Science).

Special thanks to Ruth, Molly and Jack for their support throughout.

Publisher's acknowledgements

We are grateful to the following for permission to reproduce copyright material:

Figures
Figures 6.3, 6.4, 6.10 from *Urban Drainage*, 2 ed., Spon Press (Butler, D. and Davies, J.W. 2004); Figures 6.11, 13.1 from Malcolm Blake, Coventry University

In some instances we have been unable to trace the owners of copyright material, and we would appreciate any information that would enable us to do so.

1. Introduction

This is a book about written and spoken communication, intended specifically for engineering and applied science students. Its aim is to help you learn to communicate well, and consequently to become a more successful student and a more effective practitioner.

1.1 The importance of communication skills

Our word "engineer" can be traced back to the Latin *ingenium* meaning cleverness, or natural ability. "Science" is from *scientia* meaning knowledge. The main business of professionals in engineering and applied science is to use knowledge and to be ingenious: to come up with good ideas and make them work in practice. No one works in complete isolation; there is no point in having a good idea if you are unable to communicate it. Poor communication can create ambiguity, even cause disasters. At the very least it gives a bad impression: if people think you communicate badly, they won't trust you as a professional. These people could be prospective employers, bosses, colleagues, clients, the public or the media. There's a great deal at stake: your personal career, the quality of your achievements, the benefit to society, the status and reputation of your profession – all these things depend on good communication.

Yet many graduates from technical disciplines are poor at communicating. You may be an exception of course, but employers widely believe that there is a problem. They also feel that things are getting worse. Typical comments are: "Our graduates cannot write simple letters in decent English", or "I found six spelling mistakes in three lines – I haven't time to correct other people's spelling". As a new graduate you will make progress in your career by taking responsibility, but no matter how good you are at the technical side of the job, if you can't write clearly you won't be left in charge.

Why do technical graduates tend to be poor at communicating? Let's consider what has happened to you in the last few years. By 16 you were probably showing promise in maths and science, and since then your education has been narrowly based in these areas. English was never your favourite subject and if you were taught grammar in a formal way you thought it was rather boring and have forgotten most of it by now. Am I close? Now you are on an engineering or applied science course where most of the challenge and excitement seems to be in the technical subjects.

But let's stop there. You are reading this book, which probably shows that you accept the point that technical professionals should be good at communicating. Well, now's the time for you to start improving! Most people learn better when they can see the point of what they are learning, so perhaps at this stage in your education you can get more pleasure from learning how to write well than you did when you were at school.

Engineering and applied science courses are usually very concentrated – there's plenty to learn. You probably have some guidance in communication within your course, but time constraints are such that the topic is a small element compared with the technical subjects. But don't take this as an indication of importance; remember, you will be a far more successful student if you can communicate well. You will be required to submit written work from the start of your course, and later some of your most significant assessments will be based on reports.

Improving your ability to communicate well while you are a student is something you must take responsibility for yourself. If you really want to improve, you will. You should take your communication sessions seriously, work hard at each written assignment, and use a book like this as a source of help and ideas. Finding out for yourself is an important skill for students and for working people.

English is a flexible language, constantly capable of accepting new words and expressions. English style and what is considered "correct" are changing steadily. A technical document written in the style of Shakespeare or even Sir Isaac Newton would seem strange and inappropriate to a reader today. Good English is not based on fixed and absolute rules in the way that mathematics is. Writing requires judgement, and good judgement requires confidence. You must know the "rules" before you can break them.

Of course the ability to communicate well can enhance your personal life as much as your professional life. Good communication has a good effect no matter what form it takes.

1.2 How to communicate

There is just one elementary rule in technical communication: be clear. This book is designed to help you to write and speak clearly.

When students or professionals in technical areas write or speak they should have something specific (usually factual and precise) to say. It should be possible to separate *what* is being communicated from *how* it is being communicated. If you are not sure precisely what you want to say you must stop and work it out first. When you are sure, you must know how to communicate it successfully – for the receiver to understand without loss of precision.

A characteristic of technical communication is that words are by no means the only medium. Numbers, tables, mathematical expressions, graphs, diagrams and drawings can all enhance communication. Words are only used where they are needed.

Good communication will help you become a good practitioner, but there's a lot more to being a good practitioner than simply being a good communicator. This book is certainly not intended to encourage you to develop an artificial style of communication in which how you say something is more important than what you say. The aim is to help you develop a clear style so that when you have something to say you can say it effectively. Fortunately, the technical professions are ones in which genuine achievement is more important than the appearance of success. You cannot disguise poor work by slick presentation – not for long anyway. Good communication is a means to an end, not an end in itself.

Communicating well requires effort and commitment. There are no shortcuts. However, the advice in this book should help to ensure that your effort is not wasted.

1.3 This book

This book covers the forms of communication that you are likely to use while you are a student. Your course is preparing you for your professional career, and the communication skills you learn as a student will remain of value to you when you are working. Many forms of communication – reports, letters, interviews – are as important to students as they are to professional people. However, other "professional" communication skills like writing company memos or conducting interviews are not needed much by students. These are introduced in a short chapter at the end, but are not covered in detail. There are plenty of books on professional technical communication, and some will be recommended.

Communication is essential to everyone, and the word has come to mean many things to many people. This book does not cover the technology of communications, engineering drawings (apart from diagrams in reports), management/personnel/corporate communications, advertising, public relations or media studies.

There are many ways of communicating; using language is one of them. We hope the book will be helpful to all English-speaking engineering and applied science students, including those who have not spoken English all their lives. However, the book cannot be treated as a substitute for a language textbook.

Apart from a dictionary, we hope that this is the only book on writing and speaking that you will need to use regularly while you are a student. It includes basic elements such as spelling, parts of speech, punctuation, grammar and style. These are covered in many other books – we don't claim to do the job better (though we may sometimes do it differently) – but our aim is to include all these aspects under one cover, together with their applications by students to lab reports, project reports and so on. In the later chapters we suggest other books that you might find useful for reference, and explain why.

We assume throughout that you will want to make effective use of computers for all forms of communication: word processors for text, spreadsheets for data, these or other types of software for diagrams and to support spoken presentations, and the Internet generally. There is advice throughout the book on making the most of these opportunities, but this advice is not software-specific and certainly does not attempt to take the place of a user manual.

We have already suggested that one of the reasons why graduates in technical areas are poor at communicating is that they did not think seriously enough about the principles of good English when they were at school. For that reason the early part of this book goes back to basics. Chapter 2 is about words: spelling, meaning and function. Chapter 3 is about sentences and punctuation. We hope you find this coverage useful. Even if you feel it is an insult to your intelligence, you might still consider dipping into these chapters to boost your confidence – both start with diagnostic tests.

There are several other tests in the book. They are designed to allow you to try out particular skills or put into practice some of our advice. Our answers are at the back.

The book is written in an informal style, not the sort of style that you should use when you write a formal report. Also we don't tend to preface our suggestions or advice with comments like "this is only our opinion but . . .". We tend simply to write "you should do this", or "don't do that". We hope you find the suggestions useful.

2. Words

How good is your knowledge of words? Let's start with a test.

Test 2a

1. Is the spelling of each of these words correct or not? If incorrect, give the correct spelling.

   procedure
   proceedings
   superceded
   immeadiately
   maintenance
   necessarily
   sucessfully
   occasionally
   proffessional

2. **phenomenon** – what does this word mean, what is its plural, and from which language is it derived?

3. Give two examples of *adverbs* in context.

4. What is the difference between the words in each of these pairs?

its	it's
principal	principle
affect	effect
practice	practise
imply	infer

5. Explain the meanings of the following words.

   inflammable
   literally
   verbal

---

6. More spellings. Correct or not? If incorrect, give correct spelling.

language
gauge
gaurdian
begining
parallel
develop
seperate
accomodation
committee

For answers, see page 170.

2.1 Knowledge of words

The aim of this chapter is to improve your basic knowledge of words. English words are fascinating, and the more attention you give them the more interesting they become. The relationship between spelling and pronunciation, for example, can be quite extraordinary. People who have been speaking English all their lives tend to take this for granted (but still make plenty of mistakes). People who are starting to learn English can find it mystifying.

Let's consider English words for numbers.

One, two: that's a bad start, neither is written as it is pronounced.
Three: OK.
Four: not bad, though it doesn't rhyme with flour or tour.
Five, six: OK.
Seven: not too bad, but it doesn't rhyme with even.
Eight: eh? How could anyone guess from that spelling that the word rhymes with gate?

Historically, English developed from the language of the Anglo-Saxons. It had many influences, including the incoming languages of the Vikings and the Normans. It was not until printing became common that universal spelling conventions began to emerge. Before that time language had been mostly in spoken form, with different historical influences dominant in different regions. Even while spelling conventions were becoming established, pronunciation was changing under the different influences, so that many of our spellings are effectively out of date: at one time, for example, the k in **know** was sounded.

Attempts at creating more logical spelling conventions have generally failed (though a number of words have different American and British spellings).

A good knowledge of words – their spelling, their function, their meaning, and how to use and select them – is essential for anyone who wants to communicate well. Since words are the basic components of language, improving your knowledge means going right back to basics, perhaps to some extent swallowing your pride. But it's worth it because spelling mistakes can make you look foolish. You should be sure you know how to spell **professional** before you use the word when writing about yourself; unless you are sure how to spell **superseded**, don't write it in red letters across old drawings. Of course you can check words in a dictionary if you have one handy, and that is a very good habit, but you must have a sound knowledge of spelling to know when a simple word needs to be checked. Computer spell-checks can be useful but they have severe limitations, as will be considered later. Also there are times when it is simply not practicable to use a dictionary or a computer, for example when writing out an urgent warning notice, or sending an email in a hurry.

2.2 Spelling

As discussed, the spelling of English words can be difficult and unpredictable. There are some "rules" but they all have exceptions. One that you may have heard is "i before e except after c" to which you should add "when the sound is ee", so we have:

normal:	relieve, piece, achieve
after **c**:	receive, receipt, ceiling
not ee sound:	height, weight, their
exceptions include:	protein, seize

Another useful rule is that if a consonant is doubled, the preceding vowel has a short sound (with the single consonant, a long sound) as in:

writing	written
hoping	hopping
taping	tapping
holy	holly
later	latter

So there is no excuse for: "I think I am quite good at writting English" (written by an engineering student).

The only way to improve your spelling is to take an interest in the way words are spelt. If you *care* about your spelling, it will improve.

It must be time for another test.

Test 2b

1. Is the spelling of each of these words correct or not? If incorrect, give the correct spelling.

   independant
   consciencious
   foriegn
   analyse
   diagram
   disasterous

2. Spell the words for the numbers 14 and 40.

3. Is the spelling correct or not? If incorrect, give correct spelling.

   argument
   aweful
   install
   installment
   benefitted
   anomaly

For answers, see page 10.

A list of commonly misspelt words appears on page 10. It includes all the words in Tests 2a and 2b so you can check your answers. Please check thoroughly – we don't want to be accused of perpetuating your spelling mistakes!

If you really want to work hard at your spelling, get a friend to read out these words (or record them and play them back) and see how many you can spell correctly.

Here are some other spellings to watch.

The past tense of lead ("lead from the front") is **led**; the metal with the same pronunciation is **lead.**

When you mislay something you **lose** it; the word **loose** describes a poorly tied knot.

A **programme** (American: program) is a planned sequence; the shorter spelling is used in Britain for a computer **program.**

2.3 Parts of speech

Parts of speech define the function of words. Here are some simple descriptions with examples.

Parts of speech

noun – a word that names a thing or person

student   laboratory   certificate

pronoun – a word used in the place of a noun

he   she   it

verb – a word that describes doing or being

the student *arrived* on time   *is arriving   has arrived*

adjective – a word that gives more information about a noun

a *clever* student   the *blue* folder

adverb – a word that gives more information about a verb, an adjective or a whole expression

the student spoke *cleverly*
this is *extremely* unusual
obviously this can't be allowed to continue

preposition – a word that describes a connection with a noun

you walked *into* the party   effects *of* studying

conjunction – a linking word

Paul *and* Mary
I've been to the lectures *but* I don't understand them

Commonly misspelt words

Accommodation
acknowledge
acquire
analyse
analysis
anomaly
apparent
appropriate
argument
awful

Bachelor
beginning
benefited
business

Chaos
colleague
committee
conscientious

Definitely
develop
diagram
disastrous

Environment
exaggerate

Foreign
forty
fourteen
fulfil

Gauge
government
grammar
guarantee
guard
guardian

Height

Immediately
independent
install
installation
instalment
intelligent

Language
liaise

Maintenance
manoeuvre
miscellaneous

Necessarily

Occasionally
occurred
omitted

Parallel
possess

procedure
proceedings
preceding
preferred
privilege
professional
professor

Queue

Separate
simultaneous
successfully
superseded

Unfortunately
unnecessary

Weir
weird

2.4 Words in use

Plurals

Most plurals are formed by adding s or es to the end of the word, or changing y to ies:

word	words
student	students
stress	stresses
library	libraries

Some words do not change:

sheep
cod

Some words derived from Greek or Latin use the plural form of the original language:

phenomenon	phenomena
criterion	criteria
formula	formulae
radius	radii
maximum	maxima
minimum	minima
stratum	strata

Some words form plurals in other ways:

index	indices (use this in mathematics, indexes for books)
appendix	appendices (for books and reports, appendixes in medical terminology)
matrix	matrices
axis	axes
analysis	analyses

Apostrophes

An apostrophe (') is added to a word to indicate the possessive form or to mark where letters have been omitted.

Possessive

The standard form is 's as in:

the manager's car
Paul's calculator

With a plural, where an s has already been added, the apostrophe comes after the s:

scientists' discoveries
all students' best interests

Omitted letters

When we shorten **I did not** to **I didn't** we add an apostrophe to show where a letter has been omitted. In **I can't** the apostrophe marks where two letters have been omitted, and in **I won't** it shows only some of the omissions.

Cases to watch

It's is short for **it is**, whereas the possessive pronoun **its** has no apostrophe: "It's a good design and its advantages are obvious."

Some people try to use apostrophes for unusual plurals like "throughout the 1990's", but there is no need. You can simply write:

throughout the **1990s**
all staff have **PhDs**

Capital letters

Nouns which are names or titles usually start with a capital letter:

Paul   Aberdeen   Department of Computer Science

Minor words in a long title (**of** above) are not usually given capitals.

People who are unsure about capital letters tend to use them too often. If you want to say that someone has got a job in forestry you should not use a capital. However if you are saying that you studied Forestry at university you should use a capital because it is the title of the course. In the last sentence there is no capital for university; that would only be appropriate when referring to a specific university.

Abbreviations and contractions

When Professor is shortened to Prof. it should be given a full stop. That is an abbreviation (letters chopped off the end). When Doctor is shortened to Dr it is not given a full stop. That is because it is a contraction (letters dropped out in the middle). A contraction ends with the last letter of the original word.

Many people do not bother with the full stop after an abbreviation. Others insist that it should be included. You must make your own choice, but remember two things. A full stop after a contraction (Dr, Mr, Ltd) is wrong. And it is often more elegant not to abbreviate at all (Professor, September).

Remember the difference between these two common abbreviations of Latin expressions:

e.g. means **for example**
i.e. means **that is**

Hyphens

Some words include hyphens, for example:

cross-section
self-replicating
de-icer

Many other words which you might expect to need hyphens do not:

reinforcement
semiconductor
preamplifier
coordinates
subroutine

None of this should present any problems. If you want to know if a word has a hyphen, look it up in the dictionary.

Hyphens *between* words are considered as punctuation in 3.2, page 22.

2.5 The right word

There is no point in being able to spell a word or form its plural correctly if it is the wrong word. You should own a dictionary and get into the habit of using it to check words that you are unsure of. (Many students also seem to own a thesaurus. You may find one useful; see **Further reading** at the end of this chapter.)

Test 2c

What do the following words mean?

nouns:		adjectives:	
	calibration		fallacious
	arbitration		ingenuous
	litigation		tangible
	prose		ephemeral
	enormity		fulsome
	verbosity		obsequious
	pragmatism		implicit
			explicit
			disinterested
			indifferent
verbs:	refute		comprehensive
	precede		comprehensible
			homogeneous
			heterogeneous

Answers: in your dictionary!

Words to watch

Inflammable

This means easily set on fire, from the word inflame. However, people tend to associate the in- prefix with a negative (like inappropriate). So to ensure that an important warning is not misunderstood, the words flammable and nonflammable are now common.

Literally

This word is frequently misused. You often hear people say things like "He literally blew his top". This is very hard to imagine! **Literally** means not using a figure of speech. You may be "literally out of your depth" in water, but not when your job is too hard.

Verbal

In the phrase **verbal communication**, verbal means in words, not specifically in speech. Reports and letters, as well as speeches and lectures, are forms of verbal communication. Speech can be called **oral** communication, though the word oral (meaning by mouth) seems rather anatomical; oral communication could include singing, lip-reading, painting by the mouth and kissing. **Spoken** communication might be a better term.

Difficult pairs of words

fewer/less

This should be an easy one for numerate students – fewer is for integers, less is for other quantities. So:

    less than 20%
    fewer than 50 qualified engineers

affect/effect

Affect is usually a verb, effect is usually a noun:

    the strike will **affect** deliveries
    the strike is having an **effect**

Both of the following examples from student reports are wrong.

    How does all this effect students and their chances of getting a job after graduation?
    The government's anti-inflationary policies have had a crippling affect on the construction industry.

complement/compliment

A complement is a thing that fits well or makes something complete. A compliment is a polite expression of praise. Both words can also be verbs. You would complement a team if you brought skills that the team lacked. You compliment people when you say something nice about them. The adjectives complementary and complimentary are also easy to confuse.

discrete/discreet

The first word is associated with mathematics and the second with good manners.

practice/practise

Practice is the noun, practise the verb.

I want to **practise** engineering, so I have set up a **practice** with my brother.

principal/principle

Principal is usually an adjective meaning first in importance. As a noun it can be used as the title for a head of a college. A principle is a scientific or moral law.

Among the **principles** of mechanics, students learn about **principal** stress.

plane/plain

Plane refers to a surface, real or imaginary, which is flat; plain means simple, unornamented.

relevant/relative

A common mistake is to write "facts which are **relative** to the discussion" when what is meant is "facts which are **relevant** to the discussion".

continuous/continual

Continuous means without interruption. Continual means regular but not continuous.

With **continual** distractions you get some work done. With a **continuous** distraction, none.

imply/infer

Experimental results might **imply** something about a material. You, the investigator, would **infer** that it was the case.

alternative/alternate

You may have use of the car on **alternate days** (every other day). When you don't have the car, you have to make **alternative** travel arrangements.

Test 2d

This covers all aspects of the chapter. Spot the 12 mistakes with words (don't worry about the style).

<u>Other information to support my application</u>

My principle reason for applying for this post is that I feel that C G Freeland is a Company that can be proud of it's record in innovative design. Your new activities in enviromental control (to which I beleive I can contribute) perfectly compliment your existing specialisms. I am interested in post E15 (Graduate), but would like to be considered for any alternate vacancies.

My main liesure activity is swiming. I have won county medals on no less than six ocasions, and I represented England at the last commonwealth games.

Answer on page 170.

Further reading

A dictionary

Dictionaries are surprisingly cheap to buy. You might as well have two: one pocket-sized one to carry around with you, and one medium-sized one to keep at home. The spell-checking facilities of a word processor are not a substitute.

A thesaurus

This is not essential, but many students seem to find one useful (again, cheap to buy). A thesaurus gives, for any particular word, a number of other words with similar or related meanings. There are two formats. One, usually with the title *Roget's Thesaurus*, lists words in blocks, according to meaning (each word appears once). The other has common words in alphabetical order with a list of synonyms for each.

3. Sentences

Let's start with a test.

Test 3a

Divide this piece into sentences and add other appropriate punctuation.

this is a piece of writing with no punctuation what you must do is insert it you may use commas full stops capital letters paragraphs or any other forms of punctuation that you think might be appropriate the main unit of written english is the sentence sentences can be long or they can be short a sentence really expresses one thought if the thought can be expressed in a brief statement it is quite appropriate for the sentence to be short however some thoughts are more complex and are linked together by words like since and because or but in technical english where clarity is of prime importance there is more danger in long sentences than short ones of course every sentence must contain a verb if it doesnt it is in a manner of speaking too short paragraphs are also very useful for bringing clarity to written english the break between paragraphs provides a definite pause in the text punctuation really matters because it helps to make writing clear people who do not write clearly perhaps because their punctuation is poor make a bad impression they may be excellent at their profession in other respects but if they fail to get good jobs or fail to gain their clients trust they are likely to be disappointed in their careers.

Answer on page 170.

3.1 Forming sentences

I think this test shows that punctuation helps to make sense of written words. When we speak, we give our words structure by leaving pauses, changing our tone of voice, or making gestures. Written words rely on punctuation, and the most important element is the forming of the sentences themselves.

Each sentence is a complete statement. When you are reading, you may not learn much until you have read several sentences, but you can pause for a moment after each without feeling that a statement is incomplete.

When James got up this morning

is not a sentence. The statement is incomplete; what happened when he got up this morning?

When James got up this morning he noticed that the postman had already called.

That's better. We might still like to know what was in his mail: perhaps a job offer, or a letter of rejection? But we can pause; we have read a complete sentence.

Test 3b

Here is a paragraph from a student seminar report. The main problems are in the forming of sentences. What are the mistakes, and how would you correct them (without too much rewriting)? The answer is on the opposite page, so perhaps you should cover it up.

At universities the importance of teamwork is taught by means of group assignments. Where a group is set a task with a solution that consists of many elements. The elements are divided among the group members so that each member is responsible for one particular part. This type of approach teaches students the idea of responsibility as well as how to be an active member of a team. Since each member must make a useful contribution before the group task can be a success.

Answer to Test 3b

The second and last "sentences" sound wrong because they are not complete statements. The easiest way to correct this would be to join these "sentences" to the ones before, by using commas instead of full stops.

> At universities the importance of teamwork is taught by means of group assignments, where a group is set a task with a solution that consists of many elements. The elements are divided among the group members so that each member is responsible for one particular part. This type of approach teaches students the idea of responsibility as well as how to be an active member of a team, since each member must make a useful contribution before the group task can be a success.

If we did not want to reduce the number of sentences, the original non-sentences could be adjusted to make them into sentences:

> At universities the importance of teamwork is taught by means of group assignments. In these, a group is set a task with a solution that consists of many elements. The elements are divided among the group members so that each member is responsible for one particular part. This type of approach teaches students the idea of responsibility as well as how to be an active member of a team. This is because each member must make a useful contribution before the group task can be a success.

A sentence should contain a verb. That is not a rule, it is a piece of advice. Novelists regularly write sentences without verbs:

I tried the door. Locked. What now?

But in a technical report a sentence without a verb usually gives the reader a shock.

However, the presence of a verb does not guarantee that a sentence has been written.

When I got up this morning contains a verb but is not a sentence.

Another example. (Whoops, no verb!) *Here is* another example.

Several deficiencies in the product documentation which have been identified by users.

Well, yes, but what about them? This must either continue so that a complete statement is made, or be rewritten. If it is just required as an introductory statement, the **which** can be removed.

When in doubt, keep your sentences short. Short sentences are generally suitable for communicating technical information. Don't take risks. Short, correct sentences will make you a clear communicator.

3.2 Punctuation marks

We have the following punctuation marks available to us.

1. Full stop [.]

This marks the end of a sentence.

2. Comma [,]

This is the mildest punctuation mark. It suggests that certain words should be grouped together or a slight pause taken. Here are some possibilities:

(a) a pair of commas

I've used my new calculator, the one with the special functions, to check the calculation.

(b) a pause midway

I've checked my calculations, and Dave thinks you should check yours.

(c) an early pause

However, hard work alone is not enough.

(d) a list

We have completed the calculations, the drawings, the model and the report.

Commas would always be used in (a) and (d); in (b) and (c) they are helpful rather than necessary. In (a), the one with the special functions is a separate phrase describing the calculator, and is marked off by a pair of commas. Without the commas it would be odd to see the two nouns my new calculator the one together, and someone might think that the calculator had special functions capable of checking any calculation: the one with the special functions to check the calculation. In (b) the comma saves us from momentarily thinking that two things might have been checked: my calculations and . . . In (c) the comma creates a dramatic pause and prevents any premature assumption that the sentence starts However hard . . .

In novels and newspapers, writers may economise on commas in order to step up the pace. In technical writing, clarity is the aim. If a comma helps to make your writing clear, put it in, but make sure it is in the right place. (There are more examples later.)

3. Semicolon [;]

This is a mild full stop. You might write:

I have checked the calculations. Now we can work on the model.

These are two satisfactory sentences. But they are strongly linked to each other. You could write instead:

I have checked the calculations; now we can work on the model.

So a semicolon links two statements that *could* be separate sentences into one sentence.

4. Colon [:]

A colon makes us think of a compere or master of ceremonies. "Introducing our special guest for tonight: Billy Connolly." The colon marks the moment when the compere makes extravagant arm gestures and a spotlight flashes across the stage. A colon provides an introduction.

Here is an obvious example.

You should bring the following items: boots, anorak . . .

Or:

Some things are essential: passport, traveller's cheques . . .

A colon could introduce a result or consequence.

I have checked the calculations: they are correct.

5. Paragraph

A paragraph is a group of statements. A new paragraph contains a new topic or a new line of thinking. The break between them is a definite pause. It is not wrong to write a paragraph which contains only one sentence, but if you do it frequently you may be wasting the potential of punctuation to give structure to your writing.

6. Brackets [()]

These isolate a phrase in the same way as a pair of commas (2(a)), but create a stronger separation. They are often used to suggest a change of tone: the remarks in brackets may be more personal or lighthearted (or less important). A sentence should still make sense if the expression in brackets is removed. (A whole sentence or group of sentences can also be written in brackets. In this case, the last full stop comes *inside* the closing bracket.)

7. Dash [–]

Dashes are informal – you probably put lots of them in notes or emails to friends. They usually take the place of colons or commas. They can be used in reports to create informality, but there is no situation in which they are definitely required.

8. Hyphen [-]

We have already considered hyphens within words (in 2.4, page 12). A hyphen can also be used optionally as a punctuation mark between words. Many people seem to be unaware of the usefulness of this option.

The sentence:

We have planned two day long meetings.

would be made clearer by using a hyphen:

We have planned two day-long meetings.

Here are some other examples:

long-running dispute
light-interruption sensor
radial-flow turbine

In another context we might write:

vanes create radial flow at entry

The hyphen is not there now because it is not needed to add clarity.

9. Inverted commas [" "]

These have a number of uses.

(a) Quoting directly from a source.

As Shakespeare wrote: "Life's but a walking shadow."

(b) Drawing attention to an expression which should not be taken literally. We referred to **"sentences"** in the answers to Test 3b because they looked like sentences but they had not been properly formed. In dynamic testing of a model in a laboratory you might refer to **"earthquake conditions"**. But you should be careful not to take risks with the meanings of words. Never write things like:

This "proves" that the theory is correct.

(c) Recording dialogue – not normally needed in technical writing.

10. Question mark [?]

This must be used at the end of a question.

What do these results show?

11. Exclamation mark [!]

This is not often needed when professional people write, though it may frequently be implied when they speak!

Test 3c

Simple misuses of punctuation. Identify the mistakes.

1. The number of members within a team depends on two factors; the size and complexity of the project.

2. A quality management system should be based on existing systems; amended and supplemented where necessary to conform with ISO 9001.

3. Control should be exercised throughout the whole process from start to finish, products within a subcontractor's work may have to be included.

4. What are the main problems with the current system.

Answers on page 171.

3.3 Sentences and punctuation

Let us now consider how these things fit together. We will look at some more examples, and discuss some further details.

Here are some extracts from student reports, with comments on punctuation. This is not called a test, but why not cover up the comments first and make a decision yourself?

(1) A consultant is appointed by a client to assess whether a project is appropriate, at a later date if the project goes ahead, the consultant may be involved a great deal with the technical aspects of production.

Comments: at a later date is the start of a separate statement, and should therefore be the start of a new sentence. The last comma feels as if it should be the second of a pair, isolating if the project goes ahead. So we should change this to:

A consultant is appointed by a client to assess whether a project is appropriate. At a later date, if the project goes ahead, the consultant may be involved a great deal with the technical aspects of production.

(2) It was decided after analysing the subject title, that we should concentrate our efforts on WHY it was important to know who's who in the management of a project, rather than describing, in greater detail the roles of the various members of the team.

Comments: The problems are with incomplete pairs of commas. The expressions that should be isolated are after analysing the subject title and in greater detail. That would give:

It was decided, after analysing the subject title, that we should concentrate our efforts on WHY it was important to know who's who in the management of a project, rather than describing, in greater detail, the roles of the various members of the team.

The sentence now has a lot of commas, and it would probably be worth rearranging the last part to give:

It was decided, after analysing the subject title, that we should concentrate our efforts on WHY it was important to know who's who in the management of a project, rather than describing the roles of the various members of the team in greater detail.

(3) Four months later, the contractors, reported that they were down to their last £1m. In other words, they were broke. Which stopped the banks from paying any additional cash, until they had complete assurance on the costs of completing the project.

Comments: The comma after **the contractors** is wrong. Also the last "sentence" is not appropriate. It would not make a complete statement if you read it by itself. It could be attached to the sentence before by changing the full stop after **broke** to a comma. But then the second sentence would be a complicated one, and we would have spoiled the best bit, the sentence **In other words, they were broke.** So let's keep the last sentence separate with a slight rewording. Finally the last comma is not really helpful.

> **Four months later, the contractors reported that they were down to their last £1m. In other words, they were broke. This stopped the banks from paying any additional cash until they had complete assurance on the costs of completing the project.**

(4) **We must educate people, and change their attitude to technology. If we fail to do so, there will be no one to blame, but ourselves.**

Comments: The last comma should be removed. It is possible that a politician giving a speech might leave a long pause before saying **but ourselves.** But that sort of effect cannot be achieved in writing.

Test 3d

Here are some short examples for you to sort out.

1. The electronics industry has been healthy compared with other industries this can be clearly seen in the attached graphs.

2. This, coupled with high interest rates has caused many small companies to fold.

3. Engineering will continue to be misunderstood, and we graduate engineers, are the ones who will suffer most.

4. The wall is relatively thin but, it is strengthened at regular intervals by buttress supports.

5. Although these machines rarely need maintenance, do not have regular breaks like their human counterparts and do not arrive late they were not developed to replace humans.

6. Another interesting idea, is one that is currently used in Houston Texas.

7. There is no requirement for the manager to be present, isn't this unsatisfactory.

Answers on page 171.

3.4 Further punctuation details

The last comma in a list

In a simple list like **Paul, Mary, Mark and Jo** it would be unnecessary and odd to place a comma after **Mark**. Now here is a more complicated list.

> For each overflow, I carried out a thorough survey on site, made detailed design calculations, prepared a plan and section, and supervised the completion of the contract drawings.

The comma after **section** helps to make this list clear. In a complicated list, a comma after the second-to-last item is usually a good idea.

Long quotations

Here is an imaginary quotation which explains the point.

> "These sentences have been copied word for word from a source. There could be a number of reasons for quoting them: they might form a famous passage from a book, or the quoted author might have made a point particularly clearly. The quotation is correctly given in inverted commas.
>
> "The quotation is more than one paragraph in length. Each new paragraph begins with inverted commas (to remind the reader that the quotation is continuing). It is only the last paragraph that ends with inverted commas. And these last inverted commas are placed after the full stop because the whole of the last sentence is part of the quotation."

Further reading

See Chapter 4, page 35.

4. Grammar and style

4.1 The need for judgement

When people say "That's bad English" they usually mean one of two things. Either they mean that the grammar is incorrect: that one of the rules of the English language has been broken; or they mean that the style is poor: that, for example, the writing is so wordy, or full of jargon, that it is hard to understand.

Things can be more complicated. They could be referring to something like a split infinitive, which they might insist was a point of grammar but someone else would dismiss as a matter of taste. Let's call this third category word-problems.

So good writing requires an understanding of rules, of matters of taste, and of the difficult area where the two overlap. Good taste is subjective; there are commonly held opinions but these change with time. There is no one way of writing well. Good writing, above all, requires judgement, and judgement must be based on knowledge and awareness.

This chapter contains important points about grammar, style and word-problems. It is intended to help you write good English. It may also help you defend yourself against people who criticise your English without justification.

4.2 Grammar

It is sometimes hard for people who have been speaking English all their lives to explain to those who are learning the language why a particular phrase is grammatically incorrect, and another correct. Native language speakers know how to use hundreds of principles of grammar without necessarily knowing how to define or explain them.

If this unconscious knowledge meant that English speakers never made grammatical mistakes, there would be no need for this section. In fact it is easy to make mistakes, and it is often the simplest grammatical principles that cause problems. We will consider some of these now.

Subject, object, pronouns

Subject–verb–object is a simple structure for a sentence. The subject and object are nouns or pronouns, or phrases that include them.

In **Paul hit Jim** Paul is the subject, hit is the verb, and Jim is the object. We infer this from the order of the words. **Jim hit Paul** contains the same three words but the effect is very different (on the two people).

The only words that actually change their form according to whether they are subject or object are pronouns:

I hit Jim
Jim hit *me*
They saw *us*
We saw *them*

Pronouns also change when they follow a preposition

I gave the book to *her*
She gave the book to *me*

The verb **to be** does not take an object. Consider the sentence **I am a student.** Here **a student** is called the complement. When a pronoun is the complement it has the same form as it does when it is the subject.

So when you are asked, after you have knocked on the bathroom door, "Who is it?", it is grammatically correct to answer "It is I". Of course nearly everyone actually says "It's me". When writing, however, you should try to be grammatically correct, for example:

The engineers responsible for answering queries are David and I.

People are often worried about sounding pompous when referring to themselves, and **David and me** somehow sounds more modest – but it is wrong.

In informal speech it can feel more comfortable to say things like **Dave and me will answer queries** or **me and Dave will cope with the hassle.** Here **me** is part of the subject and so is obviously ungrammatical. We would never say **me am going out for lunch.**

People who are trying hard not to say **and me** fall into the trap of saying **between you and I.** This is wrong because **between** is a preposition; it should be **between you and me.** (We would not say **I can't let you come between I and my friends.**) Similarly **leave all the queries to David and me** is correct; **to David and I** would be incorrect.

The boss has nominated David and me to deal with queries is correct because **David and me** is the object. It seems to be the **David and** that puts us off. When in doubt remove other names and work out if **I** or **me** would be correct by itself.

Here is a summary.

Correct
David and I will deal with queries (subject)
The boss has nominated David and me to deal with queries (object)
Leave all queries to David and me (after preposition)
The engineers responsible for answering queries are David and I (complement)

Singular and plural with verbs

It is easy to make a slip like:

one of the many rules of the English language have been broken.

It should be **has** because the subject is **one**.
It is also easy to write something like:

in-process inspection and testing now involves more attention to the documented quality plan.

Here the **and** makes the subject plural, so the verb should be **involve**. However:

the conduct of in-process inspection and testing now involves more attention to the documented quality plan

would be correct because the subject is singular, **the conduct**.
The following sentence is correct.

Lady Windermere, accompanied by her husband, is going to visit the plant next week.

The subject is singular: **Lady Windermere**.
It would of course be:

Lord and Lady Windermere *are* going to visit the plant.

4.3 Style

The main principle is **be clear**. In technical communication, clarity can be achieved in a number of ways. We will look at the use of numbers and diagrams in Chapter 6. Here we concentrate on clear use of words. There are no rules, only guiding principles. There is no one way of writing clearly. It is not an easy thing to do, but it is worth the effort. You must practise, and always think hard about what you write.

Here are some ideas you should bear in mind. We will list them first, then give some explanation and examples.

Style points

Always be precise
Keep it brief
Keep it simple
Be yourself
Make sure it sounds right
Make it easy to read
Be careful with: made-up words
                   metaphors
                   clichés
                   jargon

Always be precise

You should be precise when you use words just as much as when you use numbers. Technical words must be used according to their established definition. Everyday words must be selected to convey a clear meaning.

Keep it brief

If you don't like writing you may welcome the idea that you should use no more words than are needed to make your meaning clear. However, writing concisely usually takes more work than writing at length. Keeping it brief requires care and concentration. You can start by avoiding unnecessary words.

The words in italic below could simply be removed.

fewer *in number*
an *approximate* estimate
each *and every* person

It's a good idea to ration your use of **very**. It is not a precise word. You should never write that a material is very strong or a process very energy efficient. You should quantify, or at least compare with a useful reference. **Very** does not normally add anything. You may write in a letter that the delay in receiving the drawings is causing a very expensive delay, but your letter is not made more convincing by the **very** (especially if it includes more than one).

Avoid language which is rambling and vague (possibly meant to sound important).

While the ethos of quality may already permeate an organisation, management should approach the issues of quality assurance with not a little caution, while recognising the need ultimately to address them.

Students do not normally write like this, but some professional people do. If you detect any such tendencies entering your writing, try to prevent them from taking hold.

Keep it brief. Enough said.

Keep it simple

Avoid long words. Keep your sentences short. Use straightforward language. Sometimes there can be a conflict between keeping it brief and keeping it simple. In these cases, simplicity should have priority.

The software contains global input default parameter values to facilitate initial model building.

This is brief, but it is not simple. The technique of loading adjectives on to nouns (**values** has four) produces concentrated but difficult writing. Using nouns as adjectives (input, parameter and model above) has the same effect. People cannot accept information if it is delivered too quickly, so don't try to pack too much into one sentence.

Be yourself

Have the confidence to write the way you think is best. Don't copy other people's bad English. When you start work in industry, don't try to adopt a special "business style" – there is no such thing. The aim is constant: be clear.

Make sure it sounds right

Sometimes, when you check over a sentence that you have written, it seems wrong. You cannot work out why. It is grammatical, precise and simple, but it doesn't sound right. There could be all sorts of reasons: word order, for example, or too much repetition of a particular word in the piece as a whole. You are not expected to be a poet, but if something doesn't sound right to you it probably won't to your readers. You should try to find a way of improving it.

Make it easy to read

One sign that a sentence is poorly written is that it has to be read more than once to be understood. Possible remedies are improving the punctuation, adding hyphens between words (page 22) or rewording.

Made-up words

English has plenty of scope for making up words – deriving adjectives from nouns, verbs from adjectives and so on.

If, on your course, you have to study a fixed number of standard-sized **modules** (noun), then your course is **modular** (adjective). To make a course modular, someone must **modularise** it (verb). When that process is finished the course might be described as recently **modularised** (adjective). The process is **modularisation** (noun). Five years later the institution will question the benefits of **modularity** (noun: the condition, rather than the process of change).

When using these sorts of words, make sure that you use them correctly, and bear in mind that there is no need to use an ugly word when a simple expression will do. I suppose the word "modularisable", if it existed, would mean capable of being modularised – fortunately it doesn't exist.

Metaphors

A metaphor is a word used for effect in a way in which its meaning cannot be taken literally:

> there are new projects in the pipeline
> he is a giant in his field
> it was a stormy meeting

Many are so common that we tend to forget that they are metaphors: **impact, target, ceiling, kept in the dark**. This creates the danger that we use them in an inappropriate way:

> representatives are feeling snowed-under by the lack of coherent strategy.

And we can end up with some quite laughable statements:

> new water supply projects are in the pipeline
> ferry companies are taking on board new safety proposals
> it's a question of bums on seats, and we are getting our teeth into the problem

(also see Clichés, next).

Clichés

A cliché is an expression which might have been strikingly effective, even amusing, when first used, but has since become stale and worn-out. The person who first said "all we ask for is a level playing-field" probably inspired a review of procedures for competition between firms. Now the expression is badly worn and slightly irritating. Unfortunately many people, in speech at least, seem to be fond of clichés; it is almost as if they feel that these are the expressions which *should* be used.

If you care about the way you communicate you should try to avoid clichés, especially in writing. If you want to amuse, think up a fresh expression of your own. Otherwise simply say **fair competition** rather than **level playing-field**, and **rough estimate** rather than **ballpark figure**. Build up your own list of clichés to avoid. Here are some more to get you started:

moving the goal posts
at the end of the day
putting the cart before the horse

Jargon

When engineers and scientists have the opportunity to explain their work to the public they should express themselves in a way that can be understood. That's obvious. We have a reputation for communicating badly to a wider audience, and that is a great shame. If a student friend, an arts student, asks you in the pub about your final year project, don't say "Oh, you wouldn't understand"; try to explain, in everyday language, what you are doing. The low level of public knowledge of engineering and science is partly our fault.

If engineers (for example) try to communicate with the public in language that would only be understood by engineers, they are using jargon. However, when they communicate with other engineers they should make full use of the common technical language.

4.4 Word-problems

So far this chapter has given some rules for grammar, and some guidelines for style. This section considers points of English usage about which people disagree. We will give you background information, and in some cases our opinion. You must make up your own mind.

Split infinitive

An infinitive is a verb in the form: to go, to walk. It is split if a word is placed between the **to** and the verb: to boldly go, to quickly walk. Split infinitives often sound ugly, and in those cases you should avoid them. But most experts on English usage agree that it is a matter of taste not of grammar.

He or she

It used to be normal to write:

> anyone who cares about the way **he** speaks
> someone who has been speaking English all **his** life.

It was a convention; it didn't mean that the comment only applied to men.
If someone today writes

> **An engineer must have confidence in his decisions**

readers might infer, rightly or wrongly, that the writer is assuming that the
engineer is a man not a woman. It is so important not to give that impression
that this sort of wording should not be used. The problem can be solved by
writing:

> **An engineer must have confidence in his or her decisions**

But suppose the piece continued like this:

> **An engineer must have confidence in his or her decisions. He or she may have
> to defend his or her decisions to the public, to his or her employer or his or
> her client. These are the responsibilities that he or she has entrusted in him
> or her by virtue of his or her position in society.**

Things can get clumsy! It could be reworded of course, but the simplest solution
might be to put it in the plural.

> **Engineers must have confidence in their decisions**

If we are going to avoid **he** with **engineer**, we might as well be consistent. It
would be possible to change our first example to:

> **anyone who cares about the way he or she speaks**

but again we risk clumsiness. In the search for a solution, some people come
up with:

> **anyone who cares about the way they speak**

but that is ungrammatical. Why not make it plural and do the job properly?

> **people who care about the way they speak**

Data

The plural of the Latin word **datum** is data. So some people think **data** should
be treated as a plural in English:

> **the data have been analysed**

But English has other plural Latin words that are treated as singular: agenda, stamina. Moreover, in current English usage data is not the plural of datum. A datum is a fixed reference point, data is information (often in number form).

The modern way is to use data as singular:

the data has been analysed.

Informality

Contractions like **don't** and **can't** are informal. Most reports tend to be formal, and unless you want to create a particular effect you should write **do not, cannot,** etc. (This book is written informally.)

Just as formal expressions are out of place when chatting with friends, colloquial expressions are out of place in most reports. Phrases like those below do not lighten the tone, they simply stick out like a sore thumb.

stick out like a sore thumb
there is no way that . . .
it is a rip-off
the experiment was a total disaster

Further reading

Seely, John *Oxford Guide to Effective Writing and Speaking*, 2nd edition. Oxford University Press, 2005. Lots of guidance on use of English.

You may find that it is worth buying a pocket-sized book on English usage. There are several on the market, no more expensive than a small dictionary or thesaurus.

5. The writing process

5.1 Getting started

Suppose you need to write something – an important letter, a coursework report. You must stop to think before you write, so you get some paper, or switch on the computer, and think. How will you start? Think. Perhaps you will begin by explaining the problem? Hmm, think. Make a cup of coffee. Think. Close the window. Think. Think. STOP!

The problem of getting started, the fear of the blank page (or screen), is a common experience. Some people call it "writer's block", but that is a misleading expression. Fiction writers may experience writer's block – a lack of creative ideas, nothing to say. But you have something to say, otherwise there would be no point in setting out to write. The problem is putting it into words. Don't worry – the techniques are easy to master. But first remember one important thing. Everybody is different. Everyone gets ideas in a different way, thinks in a different way, and writes in a different way. Different techniques work for different people. The same person may use different approaches for different types of writing. You must work out what is best for you. The ideas that follow should help.

We're cheating with the title "Getting started", because this section is as much about finishing the writing process as starting it. But that's the point. You need to know roughly where you are aiming to go before you can start going there.

Defining the task

You should:

- define your subject precisely
- define your aim in writing about it
- define your readership.

This is not usually difficult for a short student assignment. Your subject has probably been defined for you, though this is not always the case. Your aim is to learn, and to score marks. Your readership is your lecturer or lecturers, and possibly your fellow students.

For a longer student report or a professional report this defining stage is more significant, and it is considered in detail in Chapter 9 (Reports).

Sorting out ideas

After defining the task comes an important stage: sorting out your ideas. This may be achieved by jotting down ideas, and then sorting them into a structure. Sometimes you can see the best structure immediately; sometimes it is a long struggle. Some specific methods of stimulating the development of ideas are covered in Chapter 15. Alternatively some people like to "free-write" at first, and let the structure evolve later.

Writing and improving

Fear of the blank page may simply be a case of trying to think about too many things at once: about what to say (defining the task and sorting out ideas) and how to say it (the form of words).

But there is another reason for fear of the blank page, and to understand it we must think about the *end* of the writing process rather than the beginning. When you have finished writing there will be one more important stage, in which you *improve* what you have written. So everything in your first draft can be changed later. You are not on live television, with just one chance to get it right! If you are hesitating over a blank page because you want to find the best form of words – don't bother; just use the second-best form of words, or the third-best. It really doesn't matter; it can all be improved later.

We may find that on the few occasions we are foolish enough to dither over the best form of words when we write our first draft, we take just as long improving what we have written as when we just write "rough". Sometimes "rough" writing actually produces better, more flowing sentences, and the draft takes less time to improve than a carefully written one.

In many ways, the stage in which you *write* is the least important, because you really don't have to get it right first time. And the most important is when you *improve*, because after that everything must be right.

In spite of its importance, improving should be a pleasant task; after all, the thing is written and you can relax. If you are printing out the document, try to go right away from the computer and sit somewhere comfortable. Ideally, leave it for a while and read what you've written later.

5.2 Layout

In this section we start to think about one of the most important formal communication tasks in engineering and applied science: writing reports. Several whole chapters later in the book are devoted to writing reports of different types; this section deals in particular with layout.

If you have some experience of working or studying in a technical area you will already have some knowledge of report-writing. But if most of your writing experience so far has been at school, the writing skills you have learnt are more likely to be those needed for essay-writing rather than report-writing.

An obvious difference between a report and an essay is in layout, and this has a significant effect on how the reader perceives the structure.

Report

- The structure hits you in the face
- The reader must know as quickly as possible what the report is about

Essay

- Has a structure, but the reader finds out what it is by reading

---

Here is an example of a short piece of writing in continuous prose laid out like an essay, followed by the same subject matter laid out like a report.

One benefit of taking an industrial placement is that you will find out how some of the things you are learning on the course are actually applied in practice. Your subject is practical in nature, and the sooner you spend time seeing how your knowledge and understanding can be applied to achieve something useful, the better. You will gain confidence in your abilities. It is common for students, when they start their placement, to feel anxious that they have nothing to contribute to the workplace. You should soon get over that feeling. In the early days you may be supervised closely while you develop skills and confidence, but by the end you will be making a full contribution, and may be taking significant responsibility.

The placement will give you the opportunity to develop your job-finding skills. The School offers plenty of help in finding a placement, but it is *you* who will actually win the job. You will develop skills in CV presentation and in being interviewed, and these will benefit you for the rest of your career.

All the benefits above will make you more attractive to an employer when you graduate. Some of the best vacancies are for graduates with "some experience". How can a student have industrial experience? Answer: by doing a placement. The placement may also lead directly to a job offer. A significant number of students who have impressed their employer during their placement are offered permanent jobs. That's a good feeling – knowing you have a job guaranteed. Some are offered sponsorship for the rest of their course.

Industrial placements are an optional part of the course. They last one year, and are taken after the second year of study, leading to a "Sandwich Degree". Some students worry that the discontinuity in their studies will adversely affect their results. Most students who have returned from placements say the opposite: that after a few weeks they get back into the swing of things, and find their placement has made them much more highly motivated to study and get a good degree.

Benefits of an industrial placement

You have the option of taking a one-year industrial placement after your second year of study (leading to a "Sandwich Degree"). This sheet describes the benefits.

1. Practical experience

You will find out how some of the things you are learning on the course are actually applied in practice. Your subject is practical in nature, and the sooner you spend time seeing how your knowledge and understanding can be applied to achieve something useful, the better.

2. Confidence in your abilities

It is common for students, when they start their placement, to feel anxious that they have nothing to contribute to the workplace. You should soon get over that feeling. In the early days you may be supervised closely while you develop skills and confidence, but by the end you will be making a full contribution, and may be taking significant responsibility.

3. Job-finding skills

The School offers plenty of help in finding a placement, but it is *you* who will actually win the job. You will develop skills in CV presentation and in being interviewed, and these will benefit you for the rest of your career.

4. Employability

All the benefits above will make you more attractive to an employer when you graduate. Some of the best vacancies are for graduates with "some experience". How can a student have industrial experience? Answer: by doing a placement.

5. The chance of a job offer

A significant number of students who have impressed their employer during their placement are offered permanent jobs. That's a good feeling – knowing you have a job guaranteed. Some are offered sponsorship for the rest of their course.

6. Motivation in the final year

Some students worry that the discontinuity in their studies will adversely affect their final year results. Most students who have returned from placements say the opposite: that after a few weeks they get back into the swing of things, and find their placement has made them much more highly motivated to study and get a good degree.

What devices are used in this layout to ensure that the structure hits you in the face?

- A clear title.
- Numbered subheadings with helpful titles.
- A statement at the start to say what the piece is about.

In the examples above, the writing styles and content are very similar. Sometimes essays are long winded, but reports should always be written simply. As we discussed in Chapter 1, clarity is the most important property. More detail on devices for structuring reports is given in later chapters. Even for a long and complicated report, we recommend the following test to see if the structure is clear.

If a report is good, a reader unfamiliar with its contents will . . .

- *within 10 seconds* know, basically, what it's about
- *within 30 seconds* understand how it is structured
- *within 1 minute* know its purpose.

This test is valid for student reports and for reports in industry. The test means that in industry, someone must be able to decide within a minute whether your report is worth reading.

5.3 Word processing

As we found in 5.1, a computer will play an important part in the production of your formal written work. This section is not about how to acquire word-processing skills, but gives straightforward guidance on making written work clear.

Typing

The most distracting typing mistakes tend to be with punctuation. Don't forget:

Leave a space *after* a punctuation mark, but not before.

An exception is the dash – where a space is left before and after. With brackets and inverted commas, there is no space on the "inside", but there is a space on the "outside" (unless followed by more punctuation).

Paragraphs are usually set out using one of two alternative methods.

1. In reports, it is normal to leave a line between paragraphs.
2. In books (like this one) it is normal to **indent** – move the start of each paragraph to the right – but not to leave a line.

You should not mix up the two methods. If you leave a line, don't indent. If you don't leave a line, you must indent (otherwise it is not always clear that you are starting a new paragraph). You can set up your word processor to give a consistent format.

Formatting

For the text of your document you can choose from a wide range of type sizes and fonts. The type size will have an effect on the number of pages needed, and on the ease with which the text can be read. The choice of font (Times New Roman, Arial, Courier, etc.) is to a large extent a matter of taste. You must experiment and make a decision. Be cautious about choosing an unusual font. You (or your reader) may get tired of it after a few pages.

To emphasise words you can make them **bold**, <u>underline</u> them, or use CAPITALS or *italics*. Within a sentence, bold and italics are the most common methods. There is a convention that some foreign words are printed in italics (so that the reader takes special notice, and doesn't try to read them as English words). An example might be *et al.* (meaning "and others" in a reference, explained on page 67).

An important use of emphasis is in headings. Here you have at your disposal bold, underline, capitals and italics, together with font size and the possibility of moving the title to the centre of the line. The aim is not to impress, but to ensure that the structure of your report is clear. Here are possible ways of emphasising headings.

1. INTRODUCTION	# 1. Introduction
1.1 Aims of study	## 1.1 Aims of study

The main thing is to be consistent – always use the same method of emphasising subheadings, special words and so on.

You will have control over other aspects of the printed page that can be used to make your work clear. You will be able to set the margins. Remember that space may be taken by the binding, and that if you copy your report double-sided this space will be needed in the left, then the right, margins, on alternate pages (called a "gutter margin"). You can decide whether to make the page fully justified (both left and right margins vertical lines, with the space between words on each line automatically adjusted) or left-justified (the right margin ragged, with the space between words constant). You must decide which you prefer, and then be consistent.

Headers and footers can be helpful and enhance appearance. If your report is longer than a few pages, you should use the facility for numbering pages.

Your word processor will offer ready-made "styles" for common types of documents like reports and letters.

These, and many other, options are at your service to improve your printed documents. There are generally two types of student reports in which a word-processing package has been used extensively to enhance appearance. One type comes across as showy, and says "look at me, I really know some tricks". The other is tasteful, and simply helps to make the report clear.

Checking

Remember that after you spell-check a document it is still likely to contain spelling mistakes, as in

Hear arc sum mistakes

This could arise because you confused two words and picked the wrong one, or because you made a typing error which produced, by accident, the correct spelling of a completely different word. Using the spell-checker is a very good idea, but you must also check the document carefully yourself (or ask a colleague to help you).

Saving

Remember to *save* regularly so you don't lose a lot of work in a computer failure. The machine may do this automatically at specified intervals, but if you've just finished a good bit, save anyway so you know it's all safe. Also make *backup* copies so that you have more than one saved form.

Also, try to save trees. Check your work on the screen before printing. Use *Print preview* to check the overall distribution of the text on the pages. If you can, use both sides of the paper.

6. Technical information

Students and practitioners in engineering and applied science do not communicate with words alone. In a report, technical information is presented using a *partnership* of:

numbers and symbols
tables
graphs
diagrams
... and words

These are familiar tools in a technically based course, and certainly not all of the advice in this chapter will be new to you. The aim is to bring together guidelines for using these partners, not just to record and analyse technical information, but to *communicate* it.

6.1 Numbers and symbols

How to write numbers

Small integer numbers without units are normally written as words:

three alternative designs were studied
there are **six** main reasons for proposing this scheme
students study **eight** modules in each year

Otherwise numbers are written as figures:

there are **146** possible error messages
a current of **3.7** mA
at **25** m intervals

Large numbers may be made easier to read by leaving a space between each group of three digits:

46 230   23 200 000

The traditional British practice of using a comma in this space is potentially confusing since a comma is used as a decimal point in many countries.

It is often helpful to write large or small numbers in the form:

2.69×10^6 7.03×10^{-3}

However, the choice of units (considered later) may make this unnecessary.

Accuracy

When results of measurements or calculations are given, you must be very careful about accuracy.

Strictly, 0.5 means a value between 0.45 and 0.54. 0.500 means one between 0.4995 and 0.5004.

You may, after careful consideration of accuracy, have written down a measured quantity as 2.63 (units will be discussed later). In calculation you may divide this by another measured quantity, say 4.81. Your calculator will probably give the answer as 0.5467775, but if you include all those decimal places when you quote the result in a report, you are not only being unscientific, you are also communicating badly. The result of the calculation cannot be more accurate than the measurements upon which it was based. The way you write a number communicates its accuracy. This result should be written as 0.547.

Units

You should use standard SI units and their standard abbreviations. The base SI units are:

metre	m	kelvin	K
kilogram	kg	candela	cd
second	s	mole	mol
ampere	A		

Abbreviations for other common engineering SI units are:

radian	rad	volt	V
hertz	Hz	farad	F
newton	N	ohm	Ω
pascal	Pa	siemens	S
joule	J	weber	Wb
watt	W	tesla	T
coulomb	C	henry	H

Other common units, not strictly SI, are:

angle:	degree	°
	minute	′
	second	″

hectare		ha
tonne		t
litre		l

time:	day	d
	hour	h
	minute	min

Since these are standard abbreviations, it is obviously important that you use them and do not make up abbreviations of your own. There are no plural forms; **m** is the abbreviation for metre and metres. (Maybe your Physics teacher at school used to say to the class "You are far too fond of **secs**".)

Prefixes are used for multiples of these units, for example a kilometre, km, is 10^3 metres. The common prefixes are:

	Abbreviation	Multiple
kilo	k	10^3
mega	M	10^6
giga	G	10^9
milli	m	10^{-3}
micro	μ	10^{-6}
nano	n	10^{-9}

Of course many units are made up of combinations of these units. For example, velocity is distance divided by time, metres per second. Now we hit a problem. A scientist would write this as **m s^{-1}**, but most practising engineers would write **m/s**. Some textbooks have units written one way, other books on the same subject have them written the other way. Lecturers' views also differ. While you are a student it is probably best to fit in with other people. There is no point in writing units one way when your lecturer and textbook both use the other way.

You should leave a space between the quantity and the unit:

6 mm	6.92 mm	930 kg/m^3
11 kV	0.57 m/s	25 N/mm^2

Symbols

Each branch of applied science has a fairly standard set of symbols. You will communicate most clearly if you follow the conventions. Everything you write must be consistent in use of symbols, and all the symbols that you use must be defined. If a report contains more than about ten symbols, it is worth giving a separate list of symbols in which they are all defined together. When symbols are normal letters of the alphabet (upper or lower case) it is usual to type them in *italics*. However, units should not be given in italics.

Many symbols used in applied science are Greek letters. How do you say the word for ζ or ξ? In case you've always been afraid to ask, here is a full set of Greek letters (capitals second). Now you can impress your friends with your knowledge of the Greek alphabet, *and* work out the destinations of buses when you are on holiday in Greece (though you don't have to take this book to read on the beach).

Greek alphabet

α	A	alpha	ν	N	nu
β	B	beta	ξ	Ξ	xi
γ	Γ	gamma	ο	O	omicron
δ	Δ	delta	π	Π	pi
ε	E	epsilon	ρ	P	rho
ζ	Z	zeta	σ	Σ	sigma
η	H	eta	τ	T	tau
θ	Θ	theta	υ	Y	upsilon
ι	I	iota	φ	Φ	phi
κ	K	kappa	χ	X	chi
λ	Λ	lambda	ψ	Ψ	psi
μ	M	mu	ω	Ω	omega

Equations

All equations should be written on separate lines, as clearly as possible. Each equation should be numbered in brackets on the right side of the page, and then referred to by that number.

$$H = \frac{P}{\rho g} + \frac{u^2}{2g} + z \tag{6}$$

Equation (6) is used to determine the total head . . .

If the symbols have not been defined in a separate list, each new symbol must be defined as soon as it is introduced. In this case, immediately under equation (6) we might write:

where H = total head (m)
P = pressure (N/m^2)
ρ = density (kg/m^3)
g = gravitational acceleration (9·81 m/s^2)
u = velocity (m/s)
z = vertical distance above datum (m)

In a long report which is divided into numbered sections, equations in Section 2 should be numbered (2.1), (2.2), etc.

The best way to type equations is to use the equation editor in your word processor. Without an equation editor, typed equations usually look terrible, and it would be better to write them in by hand.

Statistics

The sensible use of statistics can be of great value in making precise comments about data. Wherever you can, you should give a statistical parameter instead of using vague phrases like "reasonable agreement" or "similar to previous values". An understanding of statistics is essential when presenting technical data of any complexity. Presenting statistical information clearly and objectively can sometimes be as much a test of communication skills as of mathematical understanding.

6.2 Tables

A table is a neat way of presenting a collection of information. In an engineering report this information is often in the form of numbers. Values of each parameter are set out in the columns of the table. Each column has a heading giving a brief definition of what it contains, with a symbol and units if appropriate. Ideally tables should be kept as simple as possible, and they shouldn't contain unnecessary columns. There is no need for a column that shows an intermediate stage in a simple calculation, and certainly no need for a column in which all the numbers are the same. The column on the left may identify each row by stating, for example, the value of the independent variable, or the reference number of the specimen.

All comments in the previous section about accuracy and units apply to tables. The numbers in each column should be arranged so that decimal points are in a vertical line.

In some cases you may have a column of very large or very small numbers like this:

Volume stored
S
(m³)

5.3 × 10⁶
9.1 × 10⁶
16.8 × 10⁶
21.0 × 10⁶

Let's assume that it would be inappropriate to change the units. Writing $\times 10^6$ so many times is a nuisance and it would be convenient to write it just once, at the top of the column. You must be careful. Some people write the column heading like this:

Volume stored
$S \times 10^6$
(m³)

5.3
9.1
16.8
21.0

But this is wrong. It implies that the column contains values of *S* which have already been multiplied by 10^6 (not which still need to be multiplied by 10^6).

To convey what you really mean you should write the column heading like this:

Volume stored
S
(**m³ × 10⁶**)

5.3
9.1 etc.

This says, in effect, that each number represents so many millions of m³, which is correct. (Appropriate choice of units often means that this problem can be avoided.)

A table does not need to include many drawn lines. A horizontal line under the column headings, and lines at the top and the bottom of the table are usually all that are needed.

If there is more than one table in a report, each table must be numbered. In a long report which is divided into numbered sections, tables in Section 3 should be numbered Table 3.1, Table 3.2, etc.

Table 6.1 Experimental results

Volume of water (measured) V (m³)	Time (measured) t (s)	Discharge (V/t) Q (m³/s × 10⁻³)	Head difference (measured) H (m)	H^{1/2} (m^{1/2})
0.10	79	1.27	0.014	0.118
0.20	71	2.82	0.070	0.265
0.20	43	4.65	0.189	0.435
0.30	41	7.32	0.469	0.685
0.30	33	9.09	0.722	0.850

Table 6.1 is an example of a table of experimental results. Tables can also contain descriptive information; Table 6.2 is an example.

Table 6.2 Descriptive information

Item	Type	Supplier	Model
Valve	Butterfly (100 mm dia.)	Crane	Gem Wafer 40R series
Actuator	Pneumatic Double acting	Norbro	20RKA40
Positioner	Electro-pneumatic	IVP	K80P
Digital-to-analogue converter	12 bit, current output	3D	GPIS (modified)

6.3 Graphs

x–y graph

A common type of graph in engineering and applied science is the ordinary two-dimensional graph drawn with Cartesian (x, y) coordinates, used to investigate the relationship between two or more parameters. In mathematics, when y is a function of x, we plot y on the vertical axis and x on the horizontal axis. If, in an experiment or analysis, we are looking at the way parameter B is affected by parameter A, we plot B on the y axis, and A on the x axis.

Figure 6.1 is a simple plot of experimental results. The points are actual recorded data; the line is the trend in this data (as interpreted by the person who prepared the graph). Here, both scientific and communication skills are needed. The points communicate one thing: cold objective information; the line communicates something completely different: a judgement. You can't leave this judgement to your spreadsheet software. The computer would happily connect all the points with straight lines, or connect them with a smoothed line (Figure 6.2), but neither is necessarily appropriate. The software could help you

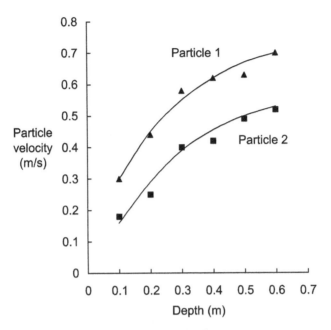

Figure 6.1 *Graph of particle velocity against depth*
x–y graph: points and lines

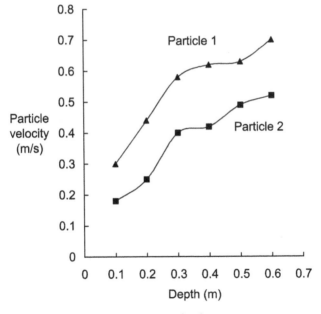

Figure 6.2 *Graph of particle velocity against depth*
x–y graph: inappropriate lines

fit a function to the points – but still you must decide which type of function is appropriate. It is important to consider whether the function should be one that passes through the origin.

Sometimes a line on a graph is used to represent something different, for example a known theoretical relationship (not a judgement). Other times it is not appropriate to have a line at all (Figure 6.3). And in other cases a line is all that is needed, when, for example, the graph is being used to present a relationship, but not specific data points (Figure 6.4).

The relative scales of the *x* and *y* axes should be established with care, so the graph is easy to understand. Some relationships need to be plotted on log scales (Figure 6.5) not linear scales.

You must make decisions about how to present your graphs; it can't be left to default settings and wizards in your package. You must take control: understand your software well, and use it to produce exactly what you want. Be careful that you are really plotting an *x–y* graph; in standard spreadsheet software a "line graph" is something different. Specify the scale and range that you really want for the axes.

Keep the presentation clear. Avoid background shading unless there is a reason for having it. Don't use unnecessary 3-D effects. Be careful with default colouring of data points (which won't be clear on a black-and-white copy).

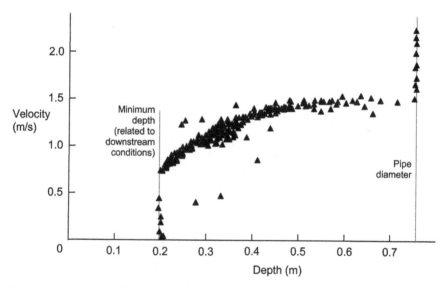

Figure 6.3 Velocity against depth scattergraph
         A scattergraph – line not appropriate

Source: © Spon Press

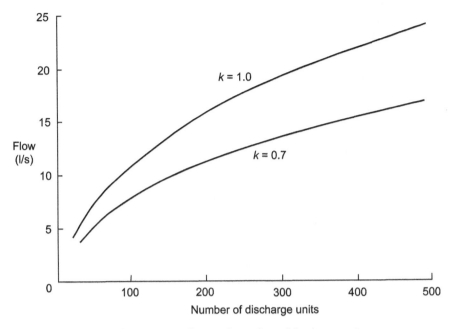

Figure 6.4 Relationship between flow and number of discharge units
x–y graph: a relationship, not specific data points

Source: © Spon Press

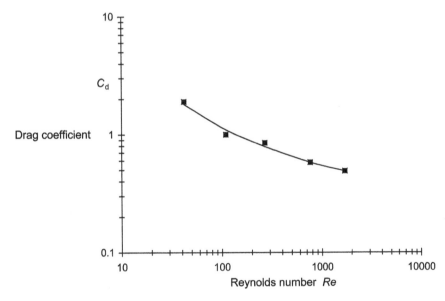

Figure 6.5 Graph of drag coefficient against Reynolds number
Log scales

Information used to label each axis should be: description of parameter, symbol (if appropriate) and units. The graph should have a number (see page 60) and a title.

The following types of graph can also be useful.

Bar chart

Bar or column charts allow simple comparison of data. They are one-dimensional graphs in which the magnitude of something is represented by the length of a horizontal or vertical bar. The bar can be divided to show how something is made up; this gives a "stacked" bar chart (Figure 6.6).

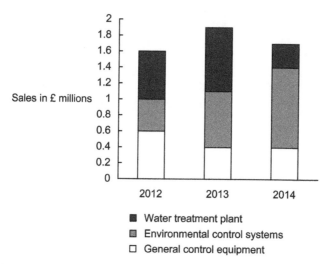

Figure 6.6 Sales figures 2012–2014
Stacked bar chart

Pie chart

A pie chart is a circle divided into segments whose areas represent relative proportions of the constituents of a whole (Figure 6.7). It communicates this information clearly and fits with popular images like "a large slice of the cake".

Histogram

A histogram can be useful for representing the general distribution of data (Figure 6.8). The y axis presents the number of data points that fall within each range defined on the x axis.

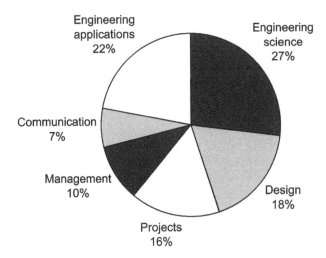

Figure 6.7 Distribution of total time on an engineering course
Pie chart

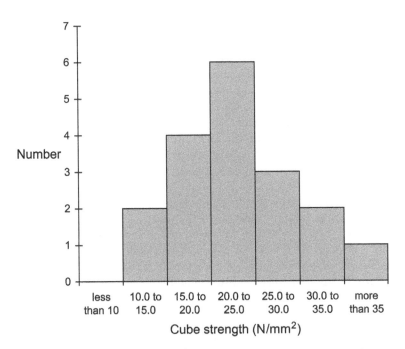

Figure 6.8 Distribution of cube strengths
Histogram

6.4 Diagrams

A diagram is a visual representation of a physical thing, system, procedure or idea.

When engineers need a visual representation of something in full detail, they produce **engineering drawings**. These are an important form of communication, but are outside the scope of this book.

A **diagram** nearly always contains an element of simplification. Simplicity and thoroughness tend to be in competition – the more you have of one, the less you have of the other. For clarity we would like our diagram to be simple, but perhaps that means we are unable to represent important details. That is where words come in. A diagram is usually most effective when used in partnership with words. Describing something with diagrams alone can be difficult (as anyone who has tried to construct an item of furniture using only the manufacturer's diagrams will appreciate). It may also be difficult to describe something in words alone. If the right diagram goes with the right words, both can be kept simple but the meaning can be clear.

Representation of physical things

A simple line diagram (Figure 6.9), in conjunction with words, can convey how a physical system operates. The diagram should show no more than is needed. If particular detail is required, a new diagram that concentrates on that aspect should be drawn.

Of course, you will regularly see more elaborate diagrams in textbooks. You may need to produce complex diagrams yourself, to show, for example, the details of a component. But your diagrams should be no more detailed than needed to provide (in conjunction with words) a clear description.

Labels should be added where they help (as in Figure 6.9). Lines pointing from the labels to the details on the diagram may be necessary, but they make the diagram look cluttered if they are not necessary. If a diagram is drawn to scale, the scale should be given. On a site plan or map, the direction *north* should be indicated.

One mistake when drawing diagrams is to try too hard to represent the *appearance* of something in addition to the information you are really trying to communicate (how it works or fits together, for example). Be clear about the purpose of each diagram. If you wish to communicate the appearance of something, it may be best to use a photograph. A good photo can save a writer a lot of words (but may not be good at showing how something works or fits together).

Representation of systems and procedures

Figure 6.10 represents a complex system. It would take several paragraphs of text to communicate the same information. Procedures for carrying out operations

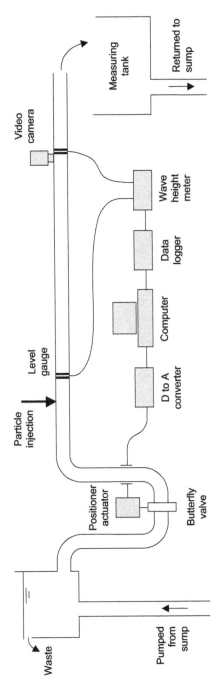

Figure 6.9 Diagram of experimental installation

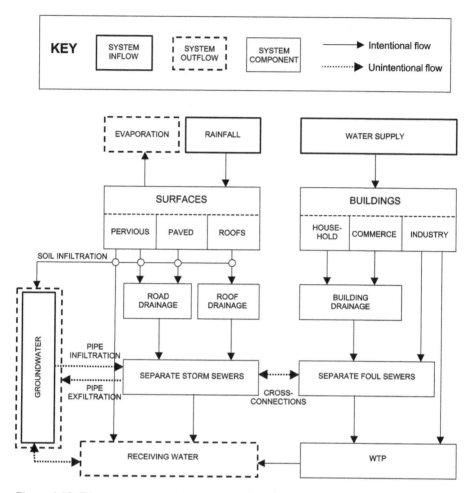

Figure 6.10 Diagram representing a system: the urban water system
Source: © Spon Press

or making decisions are also commonly defined on specific types of diagrams, including flow charts and decision trees. The best advice on designing these will come from specialist texts in your subject area.

Representation of ideas

Design ideas

Preliminary design ideas are often best represented in the form of sketches (Figure 6.11). The effective use of sketches to develop and communicate ideas is

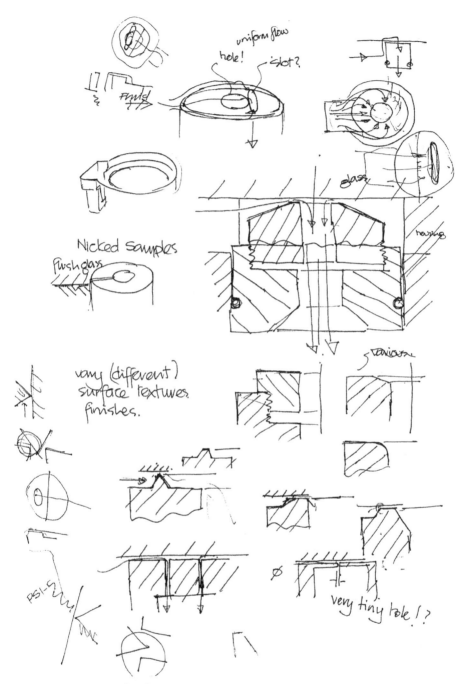

Figure 6.11 Design sketch
Source: © Malcolm Blake

a valuable skill. In an interim design report the sketches would be supplemented by description and explanation in words. If the design is taken to completion, the ideas will end up as detailed drawings.

Analytical ideas

Analytical ideas are often represented by a combination of a graph and a diagram. Figure 6.12 is a pressure diagram, showing liquid pressure increasing linearly with depth, and indicating the position of the total force.

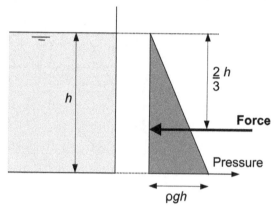

Figure 6.12 Pressure diagram

6.5 . . . and words

Where information is being presented in the form of tables, graphs or diagrams, words provide the links. They also convey the information that is best left in words. The text acts as a sort of host – introducing tables, graphs and diagrams when they are required. It follows therefore that all tables, graphs and diagrams must be referred to. If a table is not referred to from the text, it is likely to be missed by the reader, however beautiful or important it may be. Therefore all tables, graphs and diagrams must be numbered. Graphs and diagrams are usually numbered as "figures". The word **Figure** often seems to be abbreviated to **Fig.**; but there's not much point – only two characters are saved, and it's quite a short word anyway.

In a report with numbered sections, the diagrams in Section 2 should be numbered Figure 2.1, Figure 2.2, etc. We should not call this a "decimal system" since the order goes . . . 2.9, 2.10, 2.11, etc; 2.12 does not come between 2.1 and 2.2, as it would if these were real decimal numbers.

Figure 2.7, Table 5.1, etc. should start with capital letters because they are titles. Every table and figure should also have a descriptive title or caption, printed and positioned using a consistent format. A table usually has its title at the top; a figure usually has its title at the bottom.

In a report, the best position for a table, graph or diagram is as close as possible to the text which refers to it. A small diagram placed between sections of text can make a page look interesting and attractive, but it should be placed at a break between paragraphs. If more than about five pages of graphs are referred to at one point, it would be better to place them in an appendix (more about this in Chapter 9, Reports).

The results of a simple laboratory experiment may be given in a table and then plotted on a graph. However, where there is a great deal of data you should not present it all in both tables and graphs.

Here are checklists to help you ensure your tables, x–y graphs and diagrams follow good practice.

Checklist for a table

- Is it numbered?
- Does it have a title?
- Is it referred to from the text?
- Is it explained/discussed in the text?
- Is it well presented and clear?
- Does each column have an unambiguous heading? Units?
- Do all values have appropriate and consistent accuracy?
- Are all columns necessary?

Checklist for an x–y graph

- Is it numbered?
- Does it have a title?
- Is it referred to from the text?
- Is it explained/discussed in the text?
- Is it well presented and clear?
- Are the parameters on the correct axes (y against x)?
- Are axes labelled with parameter name and units?
- Is any assumed line based on sound interpretation?
- Are different points and lines distinguishable?

Checklist for a diagram

- Is it numbered?
- Does it have a title?
- Is it referred to from the text?
- Is it explained/discussed in the text?
- Is it well presented and clear?
- Does it show what is needed and no more?
- Are scale, labels and key given if needed?

Further reading

Liengme, Bernard V. *Guide to Microsoft Excel 2007 for scientists and engineers.* Academic Press, 2009. Plenty on data processing and presentation.

To learn more about technical aspects of diagrams, refer to textbooks in your subject area (rather than books on communication).

7. Source information and good practice

This chapter is about sources of information, and about how to use them, effectively and correctly, when you communicate.

The first section considers use of the Internet as part of studying, but of course the Internet is only one of many sources available to you. You also need to make use of more traditional texts, academic journals and other printed matter. This gives you access to the full range of relevant material, especially relating to academic research. It allows you to gain different perspectives, partly because the way we read a computer screen is different from the way we read a paper document. It is easy to jump from one idea to the next on a screen, especially in a Web-type document, whereas the written document demands our attention in a different manner, and this in itself is a good thing.

Your guide in all this should be your course librarian, and you should really make use of the advice and guidance materials on offer.

After considering Internet information, this chapter presents detail on referencing your sources, and ends with some advice about making sure you don't use information in a dishonest way.

7.1 Internet information

World Wide Web

The World Wide Web is a particular aspect of Internet use, involving multimedia documents with hyperlinks. Hyperlinks are an important feature of Web material. They allow a reader to move around a document at will (the principle of "hypertext") rather than reading from start to finish, and, more significantly, to follow links to other parts of the Internet. The Web is all about linking (communicating by another name), between computers, between people and between sources of information. For this reason it is important that when you are researching and browsing you keep a good record of your search. It is easy to disappear off into the distance away from your starting point. We all do this – we type a term into our search engine, then find ourselves looking at something unrelated to what we want. Most of us have butterfly minds and are attracted to the next colourful and interesting site!

You may have skimmed through this chapter and been surprised to see that it does not contain long lists of addresses of recommended Web sites. This is partly because the addresses and details of Web sites vary very rapidly. Another reason is that Web addresses are very easy to obtain anyway. This book does not contain lists of telephone numbers for these same two reasons. But perhaps more importantly, the whole make-up of browser software and the linked nature of the Web itself makes it easy to locate information without lists of addresses. All that really matters is that you know what you are looking for. When a friend recommends a film, the only thing you need to remember is the title of the film. People don't normally say, "I saw it at the Odeon in Loft Street". It's probably not showing there any more, and they assume you can find out where to see it for yourself.

Routes to information

Profession-related Web sites

Organisations in your area of interest will have Web sites that can provide useful information, ideas and links, including:

- professional institutions
- research organisations
- companies: manufacturers, consultants, contractors, suppliers
- the Intellectual Property Office – always a good source for seeing what new ideas are coming along.

The Web sites of universities (yours and others) will help you locate expertise in particular fields, and may allow you access to other sources of information like learning packages. One area that is well served by Web-based learning material is study skills, including communication.

Published material

Many journals have electronic versions on the Web and this is a trend that is growing. However, access is often restricted to members of institutions whose library has a print subscription. The contents of some British and worldwide newspapers and magazines are on the Web. Dictionaries and other reference sources are also available.

Information about information

The catalogues of the major academic and other libraries are available via the Web. Log on to your university library and you will find all sorts of useful hints and guides to help you find information. In addition to the catalogue, you will often find information about the library and its staff, as well as subject pages which may help you find information on your subject, both within the library and outside. You will find that there is a wealth of material available, and as more material is digitised the size of online repositories of information keeps increasing.

Email and discussion lists

One way of obtaining information is by communication directly with individuals – researchers, experts, practitioners, fellow students – by email. Some aspects of this are considered in Chapter 16.

Discussion forums and the like exist for all sorts of subjects and these can be extremely useful in allowing groups of individuals with common interests to exchange ideas. Finding the good ones can be a challenge so you will need to search around, share experience with fellow students and talk to your tutors.

Using Internet information

You are bound to find something useful on the Web in almost any situation. If you don't use the Web to support your studies, you are missing out. But there may be many better ways of finding out many things. Don't forget the paper sources of information that may be far more finely tuned to your needs (like lecture handouts and textbooks) or the contents of your library that may be far more authoritative or rigorous (like academic books and periodicals). A good journal will have been reviewed by other experts before it is published.

Remember that information is not the same as knowledge, and not all knowledge is useful.

As has been mentioned, material on the Web changes quickly. If you find a site that contains information you need, it is a good idea to copy or print the information straightaway, not just save the address. You might go back later and find that the site has disappeared or been changed.

The wealth of material on the Web creates new dangers of **plagiarism**, which will be discussed later in this chapter. When you quote text or present a diagram that you have downloaded, you must make the source clear. The correct method of giving a reference to Internet material is given later in section 7.2. The rapidly changing nature of the material should also be considered. Conventional printed material remains available for a reasonable length of time, and a reference to it is valid for the same length of time. In contrast, references to Web material may be out of date before your report is read. Therefore it is good practice, when you have used Web pages as sources, to attach copies to your report as an appendix or at least to ensure that you say when you viewed the material and to include a sentence of description so that it is clear to the reader what you viewed.

Perhaps the most important warning about material on the Web is this: *it may be rubbish!* A site may have been created by people who think they are experts but are not. Don't forget that anyone can contribute to some online reference sites. The late author Alan Coren used to say that he went online each day to ensure that his date of birth was always reset by a year or two and at the end of each day it had always been restored by someone who knew the truth! You must be careful when you decide whether or not a particular site is reliable. "Don't believe everything you read" applies very strongly to the Web.

7.2 References

All the information in a piece of work must come from somewhere. It could be data from your own experiments or computations, your interpretation of those results, someone else's results or interpretation, or an established principle. You may even be quoting someone else's words. Whatever the information, it must be clear to the reader where it has all come from. There are four good reasons for this.

1. *To support a statement.* To make it clear that the statement, if not based on your own evidence (as described in your text), is based on someone else's.
2. *To show how your work relates to other people's.* By demonstrating your knowledge of other work you show that you have made the most of what is available and taken care to avoid duplication.
3. *To allow your readers to find out more information by reading the publication to which you refer.* This means that you must tell them precisely how to find the publication.
4. *To acknowledge your sources.* To show that you are not pretending that someone else's ideas or words are your own.

Whenever someone else's work is involved you must make a reference. At the appropriate point in your text you give a reference code. The code allows your reader to refer to your list of references (near the back of your text). In the list you give the precise details of the publication to which you are referring.

The reference code is either a simple number written in brackets or as a superscript, or the author's surname and the year of publication.

The list of references has a different appearance for the two systems. In the number system they appear in numerical order (usually the order in which they are referred to in the text). In the name/year system (referred to as the Harvard system) they appear in alphabetical order of author's surname.

Example (number)

Good spelling may benefit engineers in their personal lives. Two-thirds of British engineers feel that if they improved their spelling they would raise their standard of living.[1] A survey by Foster[2] has shown that 80% of engineers who cannot spell "espousal" are divorced by the age of 35.

References (number)

1. James H.H. *Aspirations of the engineer*. Longman, 2010.
2. Foster G.E. The spelling of life. *Journal of Engineering Communication*, 2011, 8, 207–215.

Example (Harvard)

Good spelling may benefit engineers in their personal lives. Two-thirds of British engineers feel that if they improved their spelling they would raise their standard of living (**James, 2010**). A survey by **Foster (2011)** has shown that 80% of engineers who cannot spell "espousal" are divorced by the age of 35.

References (Harvard)

Foster G.E. (2011) The spelling of life. *Journal of Engineering Communication*, 8, 207–215.
James H.H. (2010) *Aspirations of the engineer*. Longman.

Another example using the Harvard system is in Box 11.1 on pages 106–7.

If, in the Harvard system, you refer to two publications by the same author in the same year you distinguish them by lower case letters.

(James, 2003a)
(James, 2003b)

If there are authors with the same surname, the initials are included. If there are two authors for a particular reference, both surnames are given. If there are three or more, only the first name is given, followed by *et al.* (short for *et alii*, the Latin for "and others").

(Smith D.C., 2002)
(Smith D.G., 2001)
(James and Alfson, 2000)
(James *et al.*, 2004)

The choice of system may be specified for the type of piece you are writing. The number system creates less disturbance to the flow of the text, and is probably more suitable for professional reports and short papers. The Harvard system is more common for academic reports. Whichever system you choose, stick to it rigidly. If, for example, you use the Harvard system, don't number the references as well.

The reference in the list must be precise and thorough. This is so that an interested reader can track down, from among all published material in the world, the particular publication that is referred to. Academics are rightly fussy about references, and tend to scrutinise them closely when they are assessing reports.

You must give the following information.

Journal article

author(s)
title of article
name of journal (italic or underlined)
year of publication
volume number (bold)
issue number (in brackets), if needed
page numbers

Book

author(s)
title of book (italic or underlined)
edition (if appropriate)
publisher
year of publication

Contribution in book

author(s) of contribution
title of contribution, followed by "In:"
editor(s) of book
title of book (italic or underlined)
edition (if appropriate)
publisher
year of publication
page numbers

Paper in conference proceedings

author(s) of paper
title of paper, followed by "In:"
title of conference proceedings (italic or underlined)
volume number (bold) or volume title
location of conference
year
page numbers

Report

author(s)
title (italic or underlined)
serial number
institution (name of institution, location)
year of publication

Thesis

author
title (italic or underlined)
degree for which submitted
institution, town and country if needed
year

Web material

author or organisation
title of Web site/page (italic or underlined)
Web address or "URL" (Uniform Resource Locator)
date of publication or last revision
date you accessed the site (in square brackets)

Here are some examples. Note that the position of the year depends on the system used.

Number system *(in order: journal article, book, contribution in book, paper in conference proceedings, report, thesis, Web material)*

1. Fickson D.F. and Long S.E. Effect of infiltration on quality modelling in sewer systems. *Journal of Environmental Engineering*, 2014, 56(5), 421–429.

2. Chan C.L. *Developments in manufacturing systems*, 3rd edn. Macmillan, 2012.

3. Nichols K.H. Computer monitoring of unsteady pipe-flow. In: Khan D. ed. *Computer control and monitoring of engineering processes*. Wiley, 2013, 273–291.

4. Onof P. and Adenle J.B. Practical teaching of electronics systems design. In: *Proceedings of the 5th International Conference on Engineering Education*, 2, Naples, 2013, 397–402.

5. Jones D. and Evans D. *Survey of small engineering enterprises in Wales*, Report R72. Welsh Council for Engineering, Cardiff, 2012.

6. Walters P.R. *Recycling of construction materials*. PhD thesis, University of Devon, 2014.

7. Young A. *The Industrial Links Initiative at University of Devon* http:// www.devon.ac.uk/eng/industrial/ August 2010 [accessed 20 October 2011].

Harvard *(the same examples, but in a different order because they must be listed alphabetically)*

Chan C.L. (2012) *Developments in manufacturing systems*, 3rd edn. Macmillan.

Fickson D.F. and Long S.E. (2014) Effect of infiltration on quality modelling in sewer systems. *Journal of Environmental Engineering*, 56(5), 421–429.

Jones D. and Evans D. (2012) *Survey of small engineering enterprises in Wales*, Report R72. Welsh Council for Engineering, Cardiff.

Nichols K.H. (2013) Computer monitoring of unsteady pipe-flow. In: Khan D. ed. *Computer control and monitoring of engineering processes*. Wiley, 273–291.

Onof P. and Adenle J.B. (2013) Practical teaching of electronics systems design. In: *Proceedings of the 5th International Conference on Engineering Education*, 2, Naples, 397–402.

Walters P.R. (2014) *Recycling of construction materials*. PhD thesis, University of Devon.

Young A. (2010) *The Industrial Links Initiative at University of Devon* http://www.devon.ac.uk/eng/industrial/ [accessed 20 October 2011].

Quoting

If you wish to use someone else's precise words you must enclose them in inverted commas and give the reference. Here are two examples.

> Fickson and Long (2014) point out that "in some catchments infiltration can have a significant effect on sewer-flow quality".

> "It is important that measuring instruments do not significantly disrupt the flow in pipes." (Nichols, 2013)

Quoting in this way can be effective, but there are limits. If substantial chunks of your report are not written by you, there will be less of your own work to earn marks. Quoting without inverted commas and a reference is **plagiarism** (using other people's words or ideas as if they were yours). In academic terms that is stealing, and constitutes a crime (see 7.3).

An example of a quotation within a literature review is given in Box 11.1 on page 106.

Bibliography

A list of references is precisely that – a list of publications to which you have referred in your report. A bibliography is a list of publications that you have used during your study, or that you recommend. You might write comments on the items in a bibliography. The **Further reading** recommendations in this book make up a bibliography.

7.3 Avoiding claims of academic dishonesty

We are including this section to help you to understand that, just as with any property, other people's ideas, thoughts, diagrams and research deserve to be protected. Just because you don't know the author or the researcher who thought up an idea that you are now using doesn't mean that they should not be rewarded for their thoughts. When you buy a piece of music on iTunes you are paying the artist for the copyright on their creativity; by citing an author in a text you are effectively recognising the contribution to human endeavour that someone has made.

There are two main categories of academic dishonesty in universities. The first is **plagiarism**; this is when you either carelessly or intentionally use the ideas or words of someone else and present them as your own. The second category is **collusion**; this happens when you are working with a classmate to solve a problem that you know is to be solved individually and presented as separate pieces of work, but because of time pressures you both do part of the work and then cut and paste the other piece. In both cases of academic dishonesty you are effectively risking the value of your own study and can face very serious consequences for your academic future.

Our aim is not to scare you, but simply to alert you to the dangers. A student who commits an act of academic dishonesty can be excluded from their study and prevented from ever gaining a degree – a terrible thing to happen after years of hard work, but essential to maintain the confidence in every other degree awarded.

Most institutions these days will use a piece of software to review your work. These packages are enormous online databases and every time someone submits a new piece of work it is stored for everything else subsequently submitted to be compared against. The database stores academic papers, books, student projects and dissertations, manuals, guides, Web pages and much more. The output of these packages is only indicative, but it shows how much of your work matches that of other people. There will always be some match, because you will use simple phrases that are the same as other people's, and this is fine. What is not fine is when you use long paragraphs of text that are not referenced and are clearly lifted directly from other sources.

There are other types of work that are routinely checked for originality. For example there are packages that check software code, and there are ways in which pieces of art can be confirmed as original. Also, your tutor is an expert in a particular field and has probably read the paper that you are misquoting. It is obvious to your assessor if your style changes to that of a different author – perhaps when your language moves from formal to informal, or native to non-native in tone.

There are some amusing and absolutely amazing attempts to cheat. But the serious consequences are very sad, and cause great damage to students' careers.

Here is one final piece of advice. While it is difficult to avoid, it is not a good idea to share your work with a classmate, on paper, by email, or on that laptop in the library that you leave switched on as you fetch another book. You may both end up being accused of cheating!

Further reading

Course material, and guidance from your library, on sources of information and referencing.

Neville, Colin *The complete guide to referencing and avoiding plagiarism*, 2nd edition. Open University Press, 2010.

8. Laboratory assignments

The function of a laboratory class is to allow you to learn more about a subject and to develop skills in practical investigation. One of the most important of these skills is communicating results.

Some lecturers insist on a rigid format for the presentation of lab work. Others may simply want to see results and analysis. Obviously it is important to know what is required and to try to produce it. As with all other forms of technical communication, the most important aim is to be clear.

These pages are designed to help you write clear lab assignments and reports. It is an important skill to acquire because lab reports are technical reports in miniature. Lab reports are the perfect place to start practising technical writing.

The rest of this chapter discusses the contents of a detailed report on a laboratory class. The report on a more prolonged laboratory study will have an enlarged format based on the principles of this chapter and those of Chapter 9 (Reports) and Chapter 11 (Final year project reports).

If the format for your report has not been specified but you have been asked to write a "full" report, your list of headings should be something like this:

Titles
Aims
Theory
Apparatus
Procedure
Results
Analysis
Conclusions

We will look at these headings one by one, and consider the sort of information that should be included, with tips on good practice. If you have been asked to present only part of the list, concentrate on those sections.

Titles

These are usually:

title of experiment
name(s) of student(s)
date of experiment

In a long report the titles can have a page to themselves; in a short report that is a waste of paper.

Aims

This is a brief statement of what you set out to determine, investigate, test or confirm.

Theory

This may involve one or two explanations, and the statement of key mathematical expressions. Unless you have been asked to do so, there is no point in writing out standard derivations that are given in the textbooks.

Your experiment may not be related to theory; it may for example be a standard test procedure. Some preamble to what you actually did will still be important: it will describe the purpose of the test in practice, or how the results would be used. In this case the section might be entitled **Introduction** or **Background.**

Apparatus

Alternative titles are **Equipment** or **Experimental installation.** In this section diagrams and words should be in partnership, as discussed in 6.4 (page 56). Diagrams should be as simple as possible. Their main aim will be to show how something works, not what it looks like. Clarity is more likely to be achieved by simple sections than by elaborate perspectives.

A description of instrumentation may be necessary. The section may also contain information on the design and specification of a model or test specimen.

For the diagram and description, think about the level of detail that is really necessary; you will have to stop somewhere. The general rule for full reports is that there should be enough information on apparatus and procedure for someone else to be able to repeat your experiment.

Procedure

This is the section in which you describe how you conducted the experiment. There is a well-established tradition of using the passive voice when describing experimental procedure. In the passive voice the object of a normal sentence (active voice) becomes the subject.

> *Active:* Mary unlocked the door.
> *Passive:* The door was unlocked by Mary.

Many lecturers will expect you to use the passive, and there are some good reasons for it. You might say to someone, "I measured the thickness of the specimen in four places using a vernier". That is the active voice, and the subject of the sentence is "I". Yet that subject – the person who carried out the measurement – is not important. In fact, since descriptions of experimental procedure must be precise, perhaps you ought to point out that you were working in a group of three students, and Pete took some of the readings and Laura took others. Oh no, hang on, while you were measuring the thickness of the specimen, hadn't Pete nipped out to phone his bank?

The identity of the person who carried out the procedure is, or should be, irrelevant, and the passive voice is a device for avoiding constantly making that person the subject of sentences. So we write:

The thickness of the specimen was measured in four places using a vernier.

However, the passive voice is not the only device that can be used, and the conventional style for describing procedure can be kept very simple. For example, you might have written in your notes:

> initial readings – room temperature
> atmospheric pressure

The simplest way of writing that into a sentence is:

The initial readings were room temperature and atmospheric pressure.

That is not the passive, but it is simple and appropriate, and still avoids the unnecessary human subject in the sentence. Clumsy passive voice expressions that are common in lab reports like "it was ensured that" and "it was observed that" can be avoided. Instead of "It was ensured that the apparatus was level by adjusting the foot-screws" you could write:

The apparatus was levelled by adjusting the foot-screws.

That is still passive but not so awkward.

So, we suggest that you do not describe your procedure using a "we did that" or "I did that" style. You will probably need to use the passive voice, but not for everything. Avoid the more clumsy passive voice constructions.

Be careful with the word **then** – it can be addictive. Once you use it in the description of a procedure, it is very hard to stop using it. (This was done. Then that was done. Then another thing was done. Then . . . .) Your reader will assume that you are describing the operations in sequence, so there may be no need to write **then** at all.

Finally, keep your description simple: don't write more than is needed.

Results

Refer to Chapter 6 for the best way to handle tables and graphs within a report.

If your experiment gave a reasonable number of results, it is probably best to present them in the form of a table. You must distinguish between values that were actually measured and values that were calculated from the measurements. You should include all measurements in the form in which they were recorded. Measured and calculated values do not necessarily have to be presented on separate tables; using one table may aid calculation and prevent repetition. Think carefully about accuracy; give only the appropriate number of significant figures for measured and calculated values.

Remember to place a clear title at the top of each column defining the parameter, giving the symbol (if used elsewhere in the report) and the units. Don't use a column for identical values of a constant: a column entitled "gravitational acceleration, g, m/s^2" filled with the number 9.81 is pointless and distracting. Where a column contains calculated rather than measured values, give the expression used in the calculation at the top of the column. Even if a calculation is quite involved, present it on the table; do not write out the calculations longhand. (If you feel that the calculation is too complicated to be understood from the column heading, you could write out one sample calculation in full.) A sample table of experimental results has been given as Table 6.1 (page 50).

Your observations may not all be measurements; some of your results may be qualitative not quantitative. If descriptive observations are important, they should also be given in the results section. They too can often be presented in the form of a table.

To finish this discussion of presentation of results, here's a suggestion that is as much to do with the conduct of the practical work as the presentation of the report. Try to record measurements in the laboratory *neatly*. It helps you to plan and understand what you are doing, and can save time. If your record *is* tidy, it can go straight into your report, with calculations carried out on a spreadsheet. Life is short!

Analysis

This is an important part of the report, though of course the approach to the analysis will depend on the nature of the experiment.

Analysis of the relationship between parameters is likely to involve plotting graphs (see 6.3, page 50). Interpretation of the graphs, and perhaps comparison with theory, will often be the "aim" of the experiment.

The way you describe the outcome of your analysis, for example the comparison between experimental results and theoretical relationship, is important. Expressions like "there is reasonable agreement" are vague, though there are worse, for example: "the law is verified because all the points are near the line" (even when they are not). The obvious problem is that words like "reasonable" and "near" are not precise. You should be as precise as you can. Give a maximum percentage difference, or a regression coefficient, or some other appropriate characterisation to suit the nature of the experiment.

One aspect you should comment on, even if your results are "good", is possible errors. You should identify elements of your data which you consider to be particularly affected by errors, and also identify types of error which may have a general affect on your results. The expressions "experimental error" and "human error" are unhelpful – again, be precise. What caused the error? How large was it likely to be? Anyone who has taken a measurement can, and should, form a judgement about the error that might be associated with it. Errors can be represented diagrammatically by error bars. Figure 8.1 gives an example: it is Figure 6.1 redrawn with error bars representing a possible error of ±0.04 m/s in measured values of particle velocity. (In this case errors in the measurement of depth were judged to be too small to be communicated clearly by error bars.)

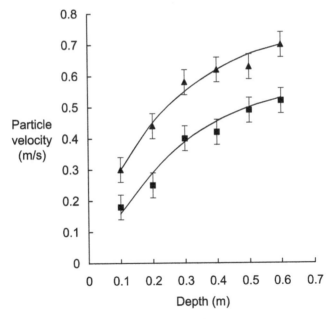

Figure 8.1 Error bars (Figure 6.1, on page 51, redrawn)

Conclusions

The real purpose of lab classes is to allow you to learn something, so perhaps the conclusions should start "I learnt the following". But in your report you have stated an aim ("to determine . . .", "to investigate . . .") and the conclusions should relate to this. You should summarise the outcome of your experiment: what you have determined, or what the investigation has shown.

Checklist for a laboratory report

- Is it the sort of report your lecturer wants?
- Have you covered all the items expected? (Depending on what you have been asked to include.)
- Have you described a clear aim?
- Could someone else carry out your experiment on the basis of your description of apparatus and procedure?
- Do your tables, graphs and diagrams comply with the guidelines in Chapter 6? (There is a separate checklist for each.)
- Are all values quoted with appropriate accuracy?
- Is it clear which of your results are measured and which are calculated from measurements?
- Have you thought about what the results really mean?
- Have you drawn conclusions that relate to your aim?

Further reading

Kirkup, Les *Experimental methods: an introduction to the analysis and presentation of data*. Wiley, 1994. A good guide to processing and presenting data from experiments.

9. Reports

This chapter considers the general skills that are needed for writing technical reports. Lab reports have already been introduced in Chapter 8. Other types of technical documentation, not usually referred to as reports, are considered in Chapter 10. The final year project report, often the most ambitious report that a student writes, is covered in Chapter 11.

Across all branches of engineering and applied science, the variety of reports written by students is wide. The properties of specific types of report are considered later in this chapter. However, the most important report-writing skills are those that are applicable to all types of report, and the main aim of this chapter is to help you acquire them.

9.1 Defining the task

A report is a formally structured collection of written information that a person or group wishes to communicate. A report is self-contained, is about something specific, and is written for a particular reason.

The important first stage in producing a report is **defining the task**. We considered this stage in Chapter 5 (The writing process). It involves:

defining the subject
defining your aim in writing about it
defining your readership

There is an element of artificiality when a student makes these definitions. The **subject** should have been set at the start of the assignment, but you may need to refine its definition. For a student there are two types of **aim**. The first is to learn and to earn marks. The second is the sort of aim that the report would have if it were being written in industry. In the case of a feasibility study report, for example, it would be to assess the appropriateness of a project. You must attempt to achieve both types of aim in the way you write the report: the second explicitly, and the first implicitly. The **readership** can also be defined in two ways: either as the assessors of the work or as the readers of such a report in practice. The report must be designed for the second readership: a feasibility study report for the client. (The first readership is addressed implicitly with every attempt to achieve quality.)

9.2 Structure

A formal and tight structure is one of the particular characteristics of a report. A novel, essay or autobiography might also be carefully structured, but the structure would be woven into the writing and not necessarily obvious to the reader. The structure of a report is made explicit: the sections are marked out by headings and subheadings, and are numbered in a systematic way.

An example list of numbered headings and subheadings for the main content of a report is given in Box 9.1.

Adaptation of twin-wire wave gauge for use in pipes

1. Introduction
    1.1 Background
    1.2 Aim and objectives

2. Twin-wire wave gauges
    2.1 General principles
    2.2 Traditional applications
    2.3 Calibration

3. Adaptation to pipes
    3.1 Design of gauge
        3.1.1 Requirements
        3.1.2 Selection from alternatives
        3.1.3 Installation details
    3.2 Data logging system

4. Practical tests
    4.1 Laboratory equipment
    4.2 Effect of gauges on flow
    4.3 Calibration
        4.3.1 General relationship
        4.3.2 Temperature effects
        4.3.3 Calibration procedures

5. Conclusions and recommendations
    5.1 General conclusions
    5.2 Recommended design

Box 9.1 Numbered headings and subheadings

This type of numbering system is generally the most suitable for a student report of any length. The headings of the main sections are numbered 1., 2., 3. Where the main sections are split, each subsection has a subheading numbered 3.1, 3.2. These subsections may themselves be split by using minor subheadings numbered 3.1.1, 3.1.2.

Some reports have a fourth layer (3.1.2.1). This makes the structure look complicated and it is generally better to avoid the need for such a subdivision by changing the overall structure.

This method of setting out structure should be incorporated into the writing process (Chapter 5) when you structure your ideas. This means that when you have sorted the ideas for your report into themes, you should establish headings, subheadings and a numbering system. This will help you to establish the clear formal structure that your report will require.

Now try out your skills at report structuring with two tests.

Test 9a

A site is being considered for the construction of an impounding reservoir. A report is to be written giving factual information on the site and its suitability. The report will not contain a comparison with other sites, or any final decision on where the reservoir should be located (this will come later).

The points that will be covered in the report are listed below *in random order*.

Create a structure for the report with headings and subheadings, and a suitable numbering system. There is no need actually to write the report; simply use the letters below to show which point will go in which section.

(a) average flow in the existing stream estimated as 100 l/s
(b) the site is not in an earthquake zone
(c) there is one rare species of wild flower that might be lost if the area is flooded
(d) recycled aggregate for construction is available from a local crushing plant for demolition waste
(e) there is some temporary accommodation locally for those working on site
(f) the site is 20 miles (32 km) from the town of Great Weezedon
(g) the area to be flooded by the reservoir is not classified as an environmentally sensitive area
(h) there are some old abandoned silver mines under the site which would need to be made safe and sealed
(i) average annual rainfall is 900 mm
(j) the area to be flooded is habitat for some protected species

(k) two farming jobs would be lost (the farmer has agreed to sell the land)

(l) there is sufficient space around the site for site offices etc.

(m) dust from construction operations must be controlled; prevailing winds would carry any dust to the nearby village

(n) the flow in the existing stream for a once-in-five-year storm is estimated as 600 l/s

(o) the ground is very suitable for dam construction

(p) the site is 3 miles (4.8 km) from the village of Little Helpington

(q) the existing stream does not carry much silt

(r) the local council has voted in favour of the scheme

(s) the local road will need strengthening to carry construction traffic

(t) the average ground level at the reservoir site is 200 m above datum

(u) exfiltration of the stored water into the ground would be low

(v) the site is located at map reference xxxxxx

(w) all excavated material will be reused within the site boundary

(x) 30 local jobs would be created once the scheme is operating through tourism

(y) water pollution caused by construction, including water pumped from excavations, must be strictly controlled

(z) the area is agricultural and in economic decline; at present there is no significant tourism in the area to be flooded

Suggested answer on page 171.

Test 9b

You are a "Party Consultant" (social parties not political ones). You charge a fee for giving advice on organisation of parties – anything from small gatherings to major events. It is a competitive field, where reputation is all-important. Your reputation depends on the quality of your reports.

You are to prepare a report on the party potential of the place where *you* currently live. Plan the structure of the report, including numbered headings and subheadings together with summaries of the content of each section. Don't write the report itself.

(You are not being asked to plan a specific party, but to assess the potential for parties. That means that by considering size of rooms, closeness of neighbours, kitchen facilities, parking, public transport etc., you should determine the most suitable type(s) of party for your place.)

Suggested answer on page 172.

9.3 Beginning and end

We have considered how to organise the main content of the report, now we must think about some important elements that you should place at the beginning and at the end. Unless the report is only a few pages long, it should have some or all of the following.

Title page
Summary
Contents page

1. Introduction

    .
    .
    .   numbered sections
    .   of the report
    .
    .

  Conclusions

References
Appendices

It is often appropriate for several other types of list to follow the contents page. These could include:

List of figures
List of tables
List of symbols
Glossary (list of definitions in alphabetical order)

Here is more detail on the main sections at the beginning and end of a report.

Title page

This should give:

  name of institution and course
  the type of assignment
  the title
  name(s)
  date

If your course has a standard cover sheet, you must of course use it.

Summary

This should be a summary of the whole report – from introduction to conclusions. A good summary is of great value to the reader. Some students are reluctant to include their conclusions in the summary. They feel that something so important should only be read in full, and then only after the main content of the report has been read. But you're not writing a detective story! You don't have to keep the outcome a secret until the last page. The convenience of the reader must come first.

The summary of a normal report should be less than one page in length. Other words for **summary** are **synopsis**, which means the same, and **abstract**, which is a summary that could be used outside the document. An example of a summary or abstract is given in Box 9.2.

An **executive summary** is longer – maybe two or three pages. It is a summary for a reader who is unlikely to be reading any other part of the report.

SUMMARY

Twin-wire wave gauges, well-established devices for monitoring wave heights in open tanks, consist of two parallel rigid wires, set vertically in the water. The conductance of the electrical circuit formed with the water varies with water depth, and therefore if a suitable calibration is applied, the output to a data logger will provide a record of varying water depth. Calibration (the relationship between water depth and conductance) is based on a series of readings of still water height and meter output. Conductance of the water itself is affected by temperature and chemical composition, and frequent recalibration is necessary.

This report describes a project in which the principle was adapted for use in pipes, by using strips of brass attached to the inside surface of the pipe. Different sizes of strips were assessed, as well as alternative configurations and spacings. The optimum design was installed in an experimental pipe system, and assessed for suitability as a practical system for recording varying water depths. It was found that the gauges did not cause excessive disturbance to the flow (no more than a typical pipe joint). The problems of calibration were investigated in some detail. Frequent recalibration was found to be necessary during the first few hours of running the equipment. After that the calibration remained constant. The most likely cause was thought to be a change in chemical composition of the water as a result of the disturbance of sediments in the laboratory sump.

It was concluded that the recommended design provides an effective and low-cost system. The main disadvantage is the need for frequent recalibration in some circumstances.

Box 9.2 Example of a summary

Contents page

The contents page sets out the structure of the report and how one part relates to another – it does more than just show where to find things. However, showing where to find things is important too, so you must remember to number the pages of the report. Pages before the main report begins, those containing the summary, list of figures, etc., can be numbered with Roman numerals.

An example contents page for a technical report by a student is given as Box 9.3.

Introduction

This is likely to be the first numbered section of the report. It will include information on the topic and on the aim, objectives and content of the study. The reader must quickly learn what the report is about.

There is a distinction between the aim and the objectives. The **aim** is the overall purpose, and the **objectives** are detailed targets. The aim of a shopping trip might be to get a new image for the summer; the objectives might be to buy new shoes, new shorts and new sunglasses. Some people say that an aim is a direction, whereas objectives are specific destinations. Objectives are achievable results, and it should be possible to say at the end of the study whether they have been achieved. It might take more judgement to decide whether the aim has been fully achieved. The aim is usually expressed in an ordinary sentence or sentences, whereas the objectives are often presented in a list, sometimes numbered, or as bullet points.

The wording of an objective can be based on the infinitive of a verb:

The objectives of the study are:
    to evaluate alternatives
    to choose the best method
    to establish a strategy for implementation

---

or on a noun:

The objectives of the study are:
    the evaluation of alternatives
    choice of the best method
    a strategy for implementation

A sample introduction to a report is given in Box 9.4.

CONTENTS

Box 9.3 Example contents page

---

1. Introduction

1.1 Background

Twin-wire wave gauges give an indication of water depth by measuring the conductance of the circuit formed between two rigid wires and the water itself. The usual application is in open wave tanks, in which the wires are set vertically. In principle the method should also be suitable for recording waves in pipes, but would need to be adapted since the wires, in their normal position, would disrupt the flow.

1.2 Aim and objectives

The aim of this study is to propose and evaluate an adaptation of the twin-wire wave gauge for use in pipes. The objectives are:

- to present the basic principles and known performance of traditional applications
- to investigate adaptation for pipes
- to evaluate alternative arrangements
- to select the most appropriate
- to evaluate its performance

Box 9.4 Sample introduction

Conclusions

A technical report in industry is likely to have the conclusions near the front, since they will be the main interest of many readers. All parts of a college report will be of interest to the assessors, so the conclusions can usually go at the end (the most logical place for a section with that name). These sorts of differences are discussed further in Chapter 19 (What next?).

The conclusions should relate closely to the aim and objectives. After you have written about both, you should check that this close relationship exists. If you wrote the objectives before you carried out the study (a very sensible thing to do), and realised later that one objective was inappropriate and could not lead to a conclusion, you can remove it from the list. If a perfectly valid objective proved impossible to achieve, you should leave it in the list, and explain in the report why it was not achieved.

There should be no fresh material in the conclusions. You should not write a conclusion about something that has not already been considered in the report.

Sample conclusions and recommendations are presented in Box 9.5.

5. Conclusions and recommendations

5.1 General conclusions

The literature indicates that twin-wire wave probes in their traditional applications in open tanks give quick response and accurate results provided they are recalibrated frequently. The recalibration is necessary because the conductance of the circuit formed with the water is related not only to the depth of the water (the parameter that the system is designed to indicate) but also to its temperature and chemical composition.

This practical study has shown that the principle can be successfully adapted for use in pipes. A variety of arrangements, based on attaching brass strips to the inside surface of the pipe, were tested. The size of the strips and their relative positions were varied to find the most suitable arrangement.

5.2 Recommended design

The most suitable design from the point of view of ease of fitting and accuracy was achieved using strips with a thickness of 0.5 mm, width 6 mm, placed vertically inside the pipe, parallel, with a 6 mm separation.

This design was installed in an experimental pipe system, and found to give accurate results for varying water depths. The gauges caused some disturbance to the water surface but no more than a typical pipe joint. This was considered acceptable.

The recommended design provides an effective method of measuring wave heights in pipes. The cost of the system is low (assuming suitable meters and data loggers are available). Frequent recalibration was necessary at certain times, thought to be a result of changes in chemical composition of the water. This was found to be the main disadvantage.

Box 9.5 Sample conclusions and recommendations

References

When other publications are referred to in the text, their details must be given in a list at the end of the report. This is treated in detail in section 7.2 (page 66).

Appendices

If detailed information on a particular aspect would break up the flow of text in the main report, it can be placed in an appendix. Examples are tables of results which have already been summarised in the report, or a series of graphs of which a representative sample has already been presented and discussed.

Checklist for format of a major report

- Is there a clear structure, with numbered headings and subheadings?
- Have you included: title page
                                   summary
                                   contents page
       if appropriate:
                                   list of figures
                                   list of tables
                                   list of symbols
                                   glossary
                                   references
                                   appendices?
- Does your summary cover the whole report from introduction to conclusions?
- Are the pages numbered?

9.4 Style

The early chapters of this book are designed to help you write good English and use a style that is appropriate for a technical report. If you want to sharpen up your skills you should refer back to the relevant chapters:

2. WORDS: spelling, use, meaning
3. SENTENCES: sentence formation, punctuation
4. GRAMMAR AND STYLE

If your English is good and your meaning is clear, you are well on the way to producing a good report.

Don't underestimate the time and effort needed to write well. Some people think they are not good at writing when really they have just never taken the time to try.

The checklist that follows identifies important aspects.

Checklist for text of a report

- Have you checked spelling by eye as well as by computer?
- Have you checked your use of words in all cases where you were uncertain?
- Do all your sentences form complete statements?
- Have you checked your use of punctuation in all cases where you were uncertain?
- Have you checked your grammar in all cases where you were uncertain?
- Is your style precise, brief and simple?
- Does it sound right?
- Is it appropriately formal?

9.5 Appearance

Chapters 5 and 6 contain information on improving the appearance of a report using a computer. After achieving the appearance you want on the screen, you should use the best quality of printer that is available for the final copy (but check the draft copy thoroughly first).

Some experts on report writing insist that the pages in a report should be single-sided not double-sided. If your report is fairly short and you don't need many copies, this may be the best way. However, if the report is long and you need a number of copies, it is much cheaper and more environmentally sensible to make double-sided copies. You should have no reservations about doing so. You should, however, try to ensure that even-numbered pages are on the left and odd-numbered pages are on the right when the report is opened (like this book).

Covers and binding

These important details depend on the length of the report and on the standard practice on your course. We suggest the following.

1–9 pages Use a standard cover sheet, if you have one for your course. If not, for a brief coursework assignment of one or two pages, barely a "report", a separate title page is not needed. Place the title information at the top of the first page. For a piece of more than a few pages, a separate title page at the front is appropriate. Place clear plastic sheets at the front and the back, and bind with a secure plastic binder. If you have any sheets bigger than A4 that must be unfolded to read, make sure they can be unfolded without the reader having to remove the binding. Don't place the whole work in a plastic wallet from which it must be removed before it can be read. And don't place separate sheets in plastic wallets if that prevents the assessor from writing comments in the margin.

10+ pages Bind with a ring comb binder, or better, with card covers at the front and back. If there is a standard design of cover you must use that.

Appearance and first impressions are important.

9.6 Types of report

Let us now consider, in a little more detail, some of the types of technical reports that students write (excluding those that are covered in other chapters).

Every type of report cannot be covered in detail. Writing specialised reports is a specialised business and needs the advice of a specialist. Also there is no single correct way of writing anything. Every report-writing task is different and every writer has a different approach.

Visit reports

You may be required to write reports on visits you make, for example to construction sites, manufacturing plants, processing installations, exhibitions or museums.

Planning the report will involve sorting out the information you have acquired into a clear structure. You should provide introductory information which assumes that your reader knows nothing about the arrangements for your visit. It may be clearest if you begin with a series of titles similar to those below.

VISIT REPORT

Location:
Purpose of visit:
Date of visit:
Those involved:

The introduction should go on to describe what was being constructed/ processed/exhibited, with a summary of what you saw.

Some of the report can be in a narrative style.

We were shown round by Mr Graham Brown, who is the Project Engineer for the consultants, D G Gammerson and Partners. He showed us a number of interesting construction details within the Control Building . . .

However, most of the report should describe what you saw, rather than how you saw it. So rather than use a narrative style throughout: "Then we were shown the air-conditioning ducts which were . . . Mr Brown explained that during installation there had been a problem with . . ." simply write:

> **The air-conditioning ducts were . . . During installation there had been a problem with . . .**

Sometimes visits come early in a course, before you have learnt much about practical engineering; however, you should still try to master basic terminology. Specialist dictionaries and textbooks can help here. You will not need to explain basic terms in your report. You should not, for example, give a detailed explanation of every common engineering material that you saw, only the more novel ones.

The purpose of your visit may be to carry out a visual inspection, or assess the condition of a structure or installation. As with any technical description, you should use clear simple language with appropriate reference to diagrams, layouts and photographs. When you carry out an assessment, you should be careful to separate factual and indisputable observations from your own judgements and recommendations.

Fieldwork reports

In many respects fieldwork reports are similar to visit reports.

The importance of good note-taking is even greater. It is best to make your notes while in the field in a hard-covered notebook, and to look after it with care. You should try to train yourself in the discipline of keeping tidy, accurate and usable records. The records may be visual, in the form of sketches or photographs. These should be accompanied by notes giving the location, direction and vertical angle of the view. If the sketch or photo is presented in the report, the same information should be reproduced, with location and direction indicated on a plan.

You may come back from a field trip with a mass of information, and the sorting out and structuring of the ideas will need care and thought. You must choose the structure which makes the report most clear and readable; it is not necessarily best to describe things in the order in which they were visited. As with a visit report, you should not assume that the reader knows the circumstances of the trip. Location, date and purpose must be made clear in the introduction.

Visual records are important. Photographs are of great value, but a diagram drawn by you may be more useful than a photograph. A diagram may not *show* more than a photo, but it may *communicate* more.

Training reports

As a student you may need to write a training report on a period of industrial experience. As a graduate you may write training reports as part of the requirements for achieving a professional qualification. Professional training reports may have a defined format, with a specified method for classifying different types of experience.

A training report will include a technical description of the projects with which you have been involved, including any special features or problems. But the emphasis must be on your own experiences. You should describe your role in the project, any special responsibilities you took on, how your role developed, any problems you encountered and how you solved them.

Different periods of experience or changes in duties or location can be shown on a table or a bar chart.

An excellent way of preparing to write a training report is to keep a diary. You may even be required to submit your diary. In any case it is good practice to keep a diary as a professional record. There is more on recording experience in Chapter 13.

Test/investigation reports

Tests and investigations include a broad range of practical technical studies, including tests in a laboratory or on site, and investigations in the field or by computer. A particular type of test report, the student laboratory report, has been considered in Chapter 8.

A test or investigation report is likely to include many of the techniques for communicating technical information covered in Chapter 6. The general structure is likely to be:

Introduction
Description
Results
Analysis
Conclusions

Each component should be clearly separated. For example, you should present all the results before starting the analysis, even if testing was carried out in more than one stage. Results should be set out in the clearest and most logical order, not necessarily the order in which they were obtained.

For a major test/investigation which forms a final year student project, advice on preparation of the report is given in Chapter 11.

Feasibility study reports

In everyday speech, the word "feasible" is used to mean something similar to "possible". However, the purpose of a practical feasibility study is not to find out if something is possible, but to find out if it is *appropriate*. A number of criteria will be used, and these vary with the discipline, but they could be technical, economic, financial, social and environmental. A feasibility study carried out internally within a firm may be particularly concerned with profit, utilisation of resources and consistency with the firm's goals. A feasibility study for a client may be concerned with value for money and environmental impact.

A typical feasibility study examines a problem or a need, proposes alternative schemes to solve or satisfy it, compares the alternatives using the full range of selected criteria, and makes recommendations.

A feasibility study report is quite different from a proposal (to be covered in the next chapter). In a proposal the writer sets out from the start to recommend a particular course of action. In a feasibility study report the writer presents objective judgements about the options.

Design reports

Design reports may be needed at various stages in the progress of a design. A report before the start of design will specify requirements. Depending on the subject, this could be called a design specification or a brief. It must be thorough and precise, as it will have a crucial influence over the development of the design. Specifications are considered in the next chapter.

In the early development of the design, reports may be needed to communicate the ideas and concepts being considered. In many disciplines the ideas are best described visually, with sketches and diagrams forming the basis of the communication. (An example of a design sketch has been given as Figure 6.11 (page 59).)

When the design is complete it will be communicated by drawings and defining documents such as a specification. The design will be based on detailed calculations which must exist in a clear, readily understood and easily checked form. Students are likely to be required to submit the complete design with all drawings, documents and calculations, together with an introduction explaining the requirements, describing the development of the concepts, justifying the choice of particular solutions, and setting out the approach to the design calculations.

Progress reports

Communication plays an important part in maintaining progress on a project. Progress reports may be quite simple, giving comparisons between actual achievement and the planned programme, and information on resources and costs. For a project of any complexity, the analysis can be carried out by a project management computer package.

A progress report should contain the appropriate level of detail. For members of the project team the precise scheduling of each activity is of interest. For a more senior manager or the client, less detail is appropriate: an overall idea of progress and costs is sufficient.

Programmes can go out of date quickly in engineering, so progress reports must be circulated regularly and frequently.

Checklist for content of a report

- Does the report have clearly stated aims?
- Have they been achieved?
- Is the reader quickly told what the report is about?
- Is the material clear, accurate and well presented?
- Have you taken appropriate opportunities to make recommendations?
- Is the report worth reading?

Further reading

Beer, David and David McMurrey *A guide to writing as an engineer*, 3rd edition. Wiley, 2009. Practical advice on different types of reports; very American.

10. Proposals, specifications and manuals

In this chapter we consider some important technical documents that are not called reports. The significance of this type of writing varies from one discipline to another. People in industry write the documents to satisfy specific needs; students write them as a preparation for their careers. For students, there may be a need for artificial realism – a member of staff playing the part of the client, for example.

This chapter can only contain general statements about these types of document, as each specific application is different. You should refer to specialised texts for more information: your lecturers are likely to make recommendations, and there are some suggestions in **Further reading** at the end of the chapter.

Many of the skills needed for report writing, covered in the previous chapter, apply to these documents. Structuring will be important, and you are almost certain to use numbered sections. Other aspects, including summary writing, arranging contents and using diagrams, also follow the same principles as reports.

10.1 Proposals

Proposals are documents designed to persuade someone to accept an idea. Their aim is to win work for your team or your organisation. Format and length vary widely depending on the type and size of the proposal. You may be asked to follow a particular format by the potential client or funding body; part of the proposal may even go on a standard application form.

The fundamental characteristic of a proposal is that it consists of two contrasting types of document subtly combined into one. On the one hand it is a cold and factual technical document, giving facts and figures which lead to an apparently objective statement of benefits. On the other hand it tries, as surely as any advertisement or sales brochure, to persuade the reader to accept ideas. Yet it must not sound like an advertisement; it must sound like an objective statement. And the facts must be accurate, not misleading.

The elements of a proposal are likely to be as follows.

Problem definition

This will only refer to problems that your proposal will solve.

Objectives

Only those that your proposal can achieve in solving the problem.

Proposed solution

This is a summary of the work you propose.

Benefits

The benefits should be specific and quantifiable. You should show how they would be measured.

Programme and resources

The programme shows when stages of the project will be completed, and also demonstrates that you have thought carefully about the practicalities of the work. A chart can be used to illustrate the programme and also to provide a visual summary of the stages of the project. A description of physical and human resources may be appropriate, including staff CVs.

Costs

These may be estimates or precise figures, depending on the nature of the proposal.

The quality of your proposal will depend to a large extent on how good you are at seeing things from the point of view of the reader, your potential client.

Your reader will certainly want the proposal to be clear and readable. The point of your proposal, and its justification, must be made early in the document. A long proposal will need a summary at the front.

10.2 Specifications

A specification is a document that specifies. People need to specify many things, and the use of the word varies from one discipline to another. They may be internal documents, or may be passed between departments or organisations.

In many disciplines, a specification is associated with an external contract, and it accompanies drawings and other contract documents. It might specify the quality of work or materials, or specify the performance standards to be achieved. When the specification is part of a contract, it may have legal significance if there is subsequent disagreement over whether the contractor has satisfied its terms.

When you are specifying standards, you must remember that higher standards are harder to meet and therefore more expensive. It is a bad thing to specify quality standards that are unnecessarily demanding.

In all cases, the requirements or standards that are specified should be measurable and achievable. It must be possible to prove that they have (or have not) been met.

When writing a specification, you must use words carefully. Terms should be defined when there is any risk of ambiguity. However, if there is no risk of ambiguity, then explanations of common terms are not needed – a specification is not usually written for a wide audience.

Specifications are cold factual documents but they must be readable. If they are hard to read, they may be misunderstood. Above all they must be unambiguous and complete.

10.3 Manuals

In industry, good manuals can be a selling point; they can increase customer satisfaction, improve standards of safety, and reduce the amount of after-sales attention required. The manual goes with the product and to an extent determines its usefulness. The product could be a whole installation, a small device or a piece of software. Parts of the manual will be descriptions of the product, and parts will be instructions for using it. The proportions will vary with the nature of the product. The writer must distinguish clearly between description and instruction, and use appropriate language for each.

It is true that user support for many engineering products, especially software, is commonly available online rather than as a printed document. But many of the principles of clear communication are common to both forms.

Considering the needs of the reader is important in all types of writing, but is probably more important for manuals than for any other type. You need to know a bit about the likely readers, including their level of technical knowledge, and their level of familiarity with the type of product. Students should clarify these details with their lecturer. At all stages of preparation, you must attempt to see the manual from the reader's point of view, and design the content to suit. It may be necessary for more than one manual to be produced in order to satisfy the needs of different readers.

You must aim to make your manual easy to use. This will be achieved through selection of material, good writing, informative diagrams and helpful layout. The layout should not be cramped, and should be designed so that diagrams are positioned conveniently, lists are clearly set out, and emphasis is given to the most important sections. A glossary may help if it is felt that readers may not be familiar with the definition of some terms.

You cannot write a manual unless you have detailed knowledge of the product. As a student, you are most likely to be asked to write a manual for a product you have developed yourself. You are not likely to be short of detailed knowledge; the extent of your knowledge may even cause problems since it may make it hard for you to see things from the point of view of the reader, who is likely to be a user without the same detailed knowledge. This makes writing the manual a demanding test of your ability to think and write clearly. One technique that could help you overcome this type of problem is to start preparing the manual while you are developing the product. You are more likely to see things from the user's point of view when you are facing the problems for the first time yourself.

Technical descriptions must be written in simple straightforward language, with reference to diagrams wherever appropriate. They should be kept separate from instructions.

Instructions tell the user what to do. The most user-friendly instructions are those that are the clearest to read, not the most informal or entertaining. Instructions must be thorough, and it is better to state the obvious than leave out something which is not obvious, especially when dealing with safety precautions. However, including too many genuinely unnecessary things will risk alienating or boring the reader.

You must think carefully about language. Let us consider an example. At a particular stage in setting up a piece of monitoring equipment, a digital display must be set to zero using keys marked UP and DOWN. This adjustment is always necessary before taking actual readings. But how should we write this part of the instructions? Here are some alternatives.

The UP and DOWN keys allow the equipment to be zeroed.

This is a description not an instruction. Also **allow the equipment to be zeroed** is not precise; how will we know when it is zeroed?

It is necessary to ensure that the digital display is set to zero by pressing the UP and DOWN keys.

This is still not an instruction.

Press the UP or DOWN keys until the digital display reads zero.

This is an instruction, but while trying to be precise the writer has created a problem. This instruction can be followed literally without achieving the desired end. What if the digital display shows a number greater than zero and the user presses the UP key . . . ?

If the digital display shows a number less than zero, press the UP key until the display shows zero. If the digital display shows a number greater than zero, press the DOWN key until the display shows zero.

This is overdoing it! Surely we can assume that the reader can imagine the effect of pressing the UP or DOWN keys.

Set the digital display to zero using the UP or DOWN key.

This is better.

If you are required to write a manual as part of a student project, don't think of it as a writing chore, an anticlimax after the main technical challenges have been overcome. This is not writing for writing's sake, it is an integral part of achieving the goal of all your efforts – making something work in practice. Students sometimes underestimate the challenge of writing a good manual for their product.

In industry the challenge is not underestimated by the most successful companies. The writing of manuals is considered to be a highly skilled task, and one which is strongly linked to the eventual success of the product.

An effective test, and an instructive experience, is to watch as someone who is unfamiliar with your product attempts to use it for the first time by following your manual.

Further reading

For specialised advice you must refer to specialist books, and these will be recommended by the lecturers on your course. The books below concentrate on good writing rather than technical content.

Beer, David and David McMurrey *A guide to writing as an engineer*, 3rd edition. Wiley, 2009. Contains advice on writing proposals and specifications.

Sides, Charles H. *How to write and present technical information*, 3rd edition. Cambridge University Press, 1999. Includes proposals and specifications.

11. Final year project reports

11.1 Final year projects and reports

In many courses the final year project is the greatest challenge. A good project brings together the main themes of the whole course, including theory, applications, specialised knowledge and design. It calls for talents that a graduate will need when entering the profession: judgement, technical understanding, originality and the ability to communicate well. Whether you do well or badly in the other parts of the course, a good final year project is a valuable achievement in its own right. It can, for example, help you find a good job, since your supervisor can praise it in a reference to a potential employer, and you can take the report (if the project is finished) to the interview.

A successful project is a personal achievement, and that is the most satisfying aspect. You will go more deeply into this topic than any other. By the end, you will know as much about it as almost anyone else. You will make great demands of your own abilities. When you have finished, *you* can take credit for the achievements.

It is not possible to generalise about the content of the project; it could be virtually anything, including research, development or design. You may have done work in a laboratory, at a computer, in a workshop, out in the field, out in industry or in the library. There may be some product: a prototype, a model, a piece of software or a set of drawings; but the main output, in which all your thoughts, experiences, frustrations and commitment are "contained", is a report. No one will know what you did, how hard you worked, how much you achieved, if it is not described in your report. Your supervisor may be able to say on your behalf "I must say this student worked very hard . . .", but it won't count for much, when the other lecturers assessing the project reply "Well, it isn't in the report". Final year projects are the ultimate proof of the importance to students of the ability to communicate well.

You will have written quite a few reports in your course before you carry out your final year project. Yet writing the project report will be an especially challenging task. It is likely to be the longest report you will have written and potentially worth the greatest number of marks. It will need to have the qualities of a good report as described in Chapter 9. It will need to be rigorous in an academic sense – with a thorough treatment of theory, an awareness of the work of others and a clear idea of what is genuinely original. The report may in fact be called a **dissertation** or even a **thesis**, but we will stick with the word report here. The advice in this chapter is to help you make it a good one.

11.2 Planning

Planning is one of the most important aspects of carrying out a project and writing the report. You must always have a clear plan which sets realistic targets and takes account of all the other demands on your time. You should start planning the report from the start of the project. Everything that you do, right from the first piece of introductory reading, needs to be represented in some way in the final report.

One of the worst mistakes is not to leave enough time for writing. This mistake can put you in a state of panic throughout the production of the report – not the best state of mind for one of the most important stages of the project.

In most final year projects, a student is supervised by a particular member of staff. It is important to make full use of your supervisor. You should regularly discuss your plan with your supervisor and start discussing the report well before it is time to start writing. You should seek comments on the draft of the first section that you write.

11.3 Contents

The general layout of your report should follow the pattern proposed in Chapter 9 (Reports). Refer back to the following sections if they are not fresh in your mind.

9.2 Structure (page 80)
9.3 Beginning and end (page 83)

Samples that could be from final year project reports are given in various parts of this book. They are listed at the end of this chapter.

Obviously the precise nature of the contents varies enormously with the subject of the project. It is not possible to predict the headings and subheadings that would be appropriate. However, the report must cover the following aspects.

Relevance of the project

You will need to give some background information about practice in your specialised area, and show how your project is relevant.

Aims and objectives

It is important to express the aims and objectives of your project briefly, clearly and precisely. This is covered under "Introduction" in 9.3, with a sample in Box 9.4.

Theory

Most projects have some theoretical basis. Any fundamental statements of theory should be presented together in one section, not introduced when needed during the description of the project.

Literature review

Alternative subheadings would be **Previous work** or **Review of relevant studies**. Someone, somewhere must have carried out work that is relevant to yours. You must describe their work and show how yours fits in. All relevant studies should be described together in this section. Later in your report, perhaps at the end of your analysis, you may wish to point out that your interpretation is similar to someone else's. But your reader should already be familiar with that person's work from your literature review. There is a sample at the end of this chapter, in Box 11.1.

What you did

This should include any rejected options or ideas that turned out to be fruitless. Projects never go smoothly – you should describe the things that went wrong as well as the things that went right. This is partly to show your assessors how much work you did, and partly to help future investigators avoid making the same mistakes.

If your project is experimental, you will include the sort of information described in Chapter 8 (Laboratory assignments). You should give particular emphasis to interesting or unusual procedures and to safety measures. There should be enough detail to allow someone else to repeat your work.

You don't always have to describe your work in the order in which it was carried out. For example, you might have tested something at one end of a physical range, then at the other end of the range, then in the middle. But the logical order for listing the tests, or presenting the results, would be ascending order of the physical parameter.

Analysis

This could be analysis of results or a critical review of what has been achieved. This may be the hardest part to write and also the most important. This is when the real thinking is needed and where the originality lies.

Conclusions and recommendations

The conclusions must be related to the aims. You should make sensible, practical recommendations. These may include areas for further study. This can create a satisfying sense of continuity: previous work leading to your work, your work leading to future work.

Conclusions and recommendations are also discussed in 9.3 (page 87), with a sample in Box 9.5.

References

In any report, and particularly an academic one, references should be handled carefully. Look again at the advice in 7.2 (page 66).

Appendix

This is a good place to put lengthy sets of data, listings of software, etc.

11.4 Writing

First of all, unless you have read it recently, look through 9.4 (page 89).

Now let us consider the particular characteristics of writing final year project reports.

Level

You should write at a level to suit someone with your general knowledge of the subject but with no specialist knowledge in the area of your project. There is no need to explain basic principles, but you must assume that your reader has no prior knowledge of your project. You are *not* writing for your supervisor.

General style

Your style should be formal and precise. Every statement you make should be justified, and nothing you write should be vague. Use tables and diagrams wherever appropriate and no more words than are needed to make your meaning clear. Give references for all material that is not your own.

Describing what you did

A writing style for describing procedure has been recommended in Chapter 8 (Laboratory assignments). The emphasis should be on what was done, not who did it.

11.5 Technical information

Presentation of technical information is likely to be an important part of an engineering project report. Refer back to Chapter 6 (Technical information) when you need to.

An aspect of final year project reports that is often disappointing is the integration of the technical data with the descriptive text. If a diagram is not referred to from the text, many readers will not see it. Most diagrams require some explanation; if you have not made the purpose of your diagram clear, there is no point in asking your readers to look at it. Remember, communication of technical information is a *partnership* between tables, graphs, diagrams and words.

Checklist for a final year project report

- For each table, graph and diagram, use the checklists in Chapter 6
- If the project includes laboratory work, use the checklist on page 78

- Use the checklist for format of a major report on page 89
- Use the checklist for text of a report on page 90
- Use the checklist for content of a report on page 95

- Have you written a good summary?
- Have you shown the relevance of your topic to practice?
- Have you included photographs where appropriate?
- Have you related your study to the published work of other people?
- Have you quoted or referred to other people's work in a clear and consistent way?
- Have you used the recommended format for references?
- Has your supervisor checked your report, and have you acted on the suggestions?

11.6 Sample extracts

To give a flavour of the appropriate style and format, fictitious samples that could be from final year project reports are presented in this book.

Summary (or **Abstract, Synopsis**)

See Box 9.2, page 84.

Contents

See Box 9.3, page 86.

Introduction (including relevance of project, aim and objectives)

See Box 9.4, page 87.

Literature review (or **Previous work, Review of relevant studies**)

A sample extract from a literature review is given in Box 11.1. Note that the references are given in this example using the Harvard system. At the end of the report the list of references will contain details of each reference using the conventions set out in 7.2 (page 66).

Literature review

Fowles (1982) carried out extensive work on models of spiral overflows and on full-scale devices, giving detailed descriptions of flow patterns, and design recommendations. In these structures, the weir was positioned on the inside of the spiral and the continuation pipe on the outside.

A different type of spiral chamber was proposed by Lodge *et al.* (1999), called a "spiral overflow with boundary spill". This design "makes use of the known tendency of spiral motion to separate solids, but uses a different configuration from earlier applications, placing the outlet pipe in the centre of the chamber floor, and weir on the outside of the spiral".

Modern descendants of the spiral overflow are called fluid dividers. A patented design, the Bossman Overflow, in which separation of solids takes place within a complex flow pattern of upward and downward helical flow, is common in the UK (Soyinka, 2003). The arrangement is shown on Figure x.x. The internal hydraulics of this device have been modelled using computational fluid dynamics software by Hornby and Self (2011), and detailed representations of liquid movement within the chamber have been simulated (Figure x.y).

Similar use of these principles has been made in other countries, for example the US "whirl chamber" (Hemingway, 1989), and the German "Hydratank" spiral separator (Mann, 2002, 2007).

> Much of the recent development has been related to specific patented devices, but research is continuing into the more general principles of devices of this type (for example, Heine and Goethe, 2011). Swift and Austen (2012) have proposed scaling protocols for physical models. Self and Hornby (2013) have studied retention efficiency for full-scale sanitary gross solids. A review of the various types of fluid dividers in use has been given by Soyinka (2013).

Box 11.1 Sample extract from a literature review (diagrams not included)

Conclusions and recommendations

See Box 9.5, page 88.

References

See 7.2, page 66.

11.7 Spoken presentations and interviews

The assessment of many final year projects also includes a spoken presentation or an interview. Advice on spoken presentations is given in Chapter 12. Interviews for projects are included in Chapter 18.

Further reading

Any handbook or guide given out on your course to help with your final year project. (Read it carefully to make sure you know what's expected and how your work will be assessed.)

Breach, Mark *Dissertation writing for engineers and scientists – student edition.* Prentice Hall, 2009. This book gives brief and helpful introductions to all aspects.

Naoum, S.G. *Dissertation research and writing for construction students*, 2nd edition. Butterworth-Heinemann, 2007. A thorough book for students in construction or any subject in which the final year project is likely to include a survey and data analysis.

12. Spoken presentations

Most students give spoken presentations during their courses. Likely occasions are seminars, and presentations on projects. The audience usually consists of fellow students and members of staff, and sometimes invited guests.

Many graduates say this is a common activity at work, so acquiring skills as a student should be very valuable. You probably have little experience of this sometimes daunting activity, so this chapter is designed to give you plenty of help. It covers your state of mind, visual aids, and preparing and giving the presentation.

12.1 State of mind

Good English is important when giving a presentation just as it is when writing, but the main problem with speaking is not planning what to say, but managing to say what you have planned once the audience is in front of you.

Most engineering students realise that spoken presentation skills are worth developing, but very few look forward to the opportunity. The main reason is a general feeling of apprehension, which tends to be called "nerves".

There is nothing wrong with being nervous. Professional performers do not try to eliminate the feeling; it carries out the important function of focusing energy and determination. However, nerves can be a nuisance if they affect your memory or voice or manual dexterity, and that (the nuisance not the nerves) needs to be overcome.

There are techniques for controlling nerves: relaxation exercises, conscious breathing, smiling at yourself in a mirror. If something works well for you, then, of course, you should use it. But there is only one truly effective way of overcoming the effects of nerves – **preparation**. Preparation includes planning and designing content, and practising. Every good presentation is a well-prepared presentation.

Some books give advice on how to feel confident before giving a presentation. There is even advice on how to *appear* confident (for use even when you don't feel confident). Perhaps this can work for some people, but many would agree that it is impossible to convince an audience that you feel confident when you don't. Often when we pretend to be confident, it does more harm than good.

There are probably two types of nerves, one good and one bad, and the difference between the two is thorough preparation. Don't worry about "being confident"; aim instead to be **positive**, as follows:

Positive	Not positive
Nervous = excited	Nervous = worried
Enthusiasm for what you are going to say – it has all been thoroughly prepared	Concern – you have no idea how it will turn out
Confidence that everything is well planned – you've tried to think of everything	You are hoping for the best – you'd rather not think about it too much

12.2 Visual aids

Think of the television news. Even the most straightforward comparison is illustrated with some kind of visual image. The value of the pound has risen against the dollar by 0.1 of a cent. That is an easy statement to understand, and yet a simple illustration helps. We understand better when we see things as well as hear them. Visual aids can also help you in your presentation by:

emphasising your points
creating variety
taking the pressure off you, by causing the audience to look at the screen
   instead.

Most formal presentations given by students are supported by projected visual material using presentation software. In some particular circumstances an alternative is a simpler method, such as a poster. More flexible means, such as a board or flip chart, may be available.

Presentation software

Computer-driven presentations don't just look good, they offer the opportunity for integrating text, graphics, photographs and other media, and the software you use will help you to structure and deliver your presentation.

One of the main benefits is that you are forced to plan carefully (and helped considerably in the process). That, and confidence in the visual quality of the material, can certainly help to make students feel positive about giving a presentation.

Much of this chapter is concerned with making the most of presentation software. Of course, you should check that the equipment and software you need will be available before preparing your material. Also for this type of presentation it is particularly important that you practise, understand the technology, and check arrangements in the room before it is time to start. It may be necessary to control the light in a bright room.

Posters

A poster is not a direct alternative to a presentation using a computer. Posters are normally presented either to a small group or as part of an exhibition in which people are walking round looking at a number of posters. You may be giving a prepared presentation, in which case you will be effectively taking the small audience through the material on the poster. Or you may simply be responding to questions from passers-by.

A well-designed poster gives a clear message. It must be interesting visually – "eye-catching" if possible. Text needs to be large enough to be read by someone standing at least 1 metre away. In any case text needs to be brief – just the essential points. Visual material must be strong. The poster shouldn't look cluttered, and the order in which you would like the material to be viewed should be clear. An image can be used as a background for the whole poster, provided it does not clash with the main content. Posters can be designed and produced using presentation software.

Boards

There will probably be a vertical surface for you to write on, although using it may not be a good idea, as we will discuss shortly. You are most likely to use the board when you are answering questions. If it is a **white board** you must ensure that you use the correct pens. An alternative is a **flip chart**, consisting of large paper sheets which can be flipped over when full. A flip chart is not likely to be used in a room which contains a board unless there is some need to refer back through the sheets.

12.3 Preparing

Notes

If you become a cabinet minister or a celebrity you may read speeches word for word. This may be because it is crucial that you do not risk saying a single wrong word, or because someone else has written the speech for you. You will by then have mastered the difficult art of reading out a whole piece and still making it sound interesting and sincere. Alternatively if you have to give a speech at a wedding, or a speech of thanks for an important award, you may memorise every word of what you plan to say.

For a student presentation neither reading nor memorising is likely to be appropriate. Reading is not appropriate because your audience will quickly become bored unless you are an expert at reading. It is very important that you look at the audience while you are speaking, and if you are reading from a script you cannot be looking up at the same time. Memorising every word is not appropriate because your delivery will sound unnatural, and you will run the risk of disaster in the event of memory loss.

So your presentation must be based on notes of what you plan to say. Your notes must be detailed enough to prevent you from leaving out something important, but brief enough to allow you to look at the audience while you are speaking. Your notes must be prepared with great care – they will be your main support during the presentation.

You are also likely to illustrate your presentation with slides (via presentation software) giving subheadings and keywords. Some people combine this material with their personal notes (after all, both are mostly made up of subheadings and keywords). They do not use paper for their notes, they give their slides the dual function of reminding them what to say and of acting as a visual aid for the audience. This is a method well worth considering for student presentations. For the method to be successful, the material on the slides must be sufficiently detailed, and your practice sufficiently thorough, for you to be reminded of everything you want to say (even when you are nervous). Figure 12.1 gives a sample slide of subheadings and keywords, together with what might actually be said during that part of the presentation.

Preparing content

Plan your content using the "writing process" approach described in Chapter 5. Think carefully about the beginning and end. It is a good idea to prepare an introductory slide giving your name and the title of your presentation. Even if someone introduces your presentation and gives your name and subject, it is still worth showing this first slide. You will say something like "here again is the title of my presentation". There is nothing wrong with the repetition; the audience may not have taken it in the first time.

Think about what your audience needs to know. Who will they be exactly? What will they find most interesting? Remember: you know something that they don't – that's why you are giving the presentation. In your introduction, give an outline of the structure of what you are going to say.

Think of a positive way of ending. At the very least say: "That concludes my presentation. Thank you." The worst way of ending is a pause during which you realise there is nothing more to say, followed by "... well ... that's it".

Problems – CALIBRATION

◆ Problem with traditional probes

◆ Conductance of circuit

◆ Conductance v. depth

◆ Temperature, chemical composition

◆ Frequent recalibration

◆ Variation during the day – reasons?

(Spoken:)

One of the main problems with the water depth gauges is calibration. This is also a problem with the traditional twin-wire wave probes. The gauges themselves measure the conductance of the circuit formed with the water, and the calibration of a gauge is really the relationship between conductance and water depth. This is found by taking readings at a series of steady water depths.

The conductance of the water itself is affected by temperature and chemical composition. If these vary during a series of tests (as they usually do), the calibration of the gauges will become inaccurate. The gauges therefore need frequent recalibration, and this is time consuming.

This was a problem with the gauges in my experimental set-up. The time period between recalibrations varied during the day. In the morning of a typical day, three or four recalibrations were needed – even if the equipment had been run for the first hour without taking any readings. In the afternoon, recalibration was not normally necessary.

I tried to investigate the reasons for the morning variation. During one morning I measured the temperature of the incoming water and found that it did not change significantly, even though the gauges needed to be recalibrated as normal. I have not been able to work out the reasons for this. I think the most likely explanation is that the chemical composition of the water in the sump varies because deposits are being disturbed.

Figure 12.1 Keywords on a slide, and what might be said

Planning visual aids

Plan what you will say about every piece of visual material in your presentation. When showing photographs, be selective about which to show. Don't include a photo that does not show anything new, or about which you have nothing to say. Prepare all your visual material in advance. If you try to write on the board during your presentation, you will cause unnecessary delay and your writing may be hurried, nervous and hard to read.

Practising

Try to practise as much as you can. You can practise at home, speaking to yourself while looking at the slides on your computer. Think about how long each slide should be shown. Time the whole presentation and try to pace yourself. You can also practise in front of a mirror, or you could try to persuade some friends to be your audience so you can give the whole presentation to them.

Try also to practise in the room in which the presentation will be given (or a similar one). Find a position at the front of the room where you are comfortable and which allows the audience to see the screen. Practise using the computer projector or other visual aids. Make sure all software needed is installed, and think about how you will operate it, including any switching between one application and another. Go to the back of the room to get an idea of how your presentation will look from there. If a remote control is available, make sure you know how to use it.

When the big day arrives, get to the room well before the session starts and have one more think about where to put things and where to stand. Check software once more and think about how you will load your presentation. If you are using a separate DVD player, make sure you know how to switch to it, and set the volume at the appropriate level.

You must of course have your presentation with you, and it's a very good idea to have a back-up copy. Ideally these should be on different types of storage device.

12.4 Making best use of presentation software

Here are three guiding principles.

1. You are already an expert

Whether or not you are familiar with using presentation software yourself, you are already an expert because you almost certainly have plenty of experience of being on the receiving end. Think about some of the good ways you have seen the potential of the software used, by your lecturers in classes, or on other occasions. Steal the best ideas. Recall the times when computer presentations have been boring, and identify approaches you will definitely avoid.

2. Concentrate on what you want to achieve, not on what the software can do

Start by assuming that the software will do what you want, and plan the presentation you really want to give. Finding out how to use the software is the easy bit. If a friend describes some amazing thing you can do, make use of the idea if you think it will help you achieve what you want, but equally be prepared to say (or think) "so what?". The key to making best use of presentation software is to make sure the technology is enhancing your presentation, not constraining it. This requires a conscious decision – real judgement about how you want your presentation to be designed. Remember that you don't have to use software to support the whole presentation.

The choice of designs is beguiling, but be tasteful rather than try to impress. Use two or three colours only – don't overdo it. The facility for calling up bullet points one by one is a helpful device for giving structure to your presentation. But the gimmicky ways in which bullet points can be made to appear on the screen can be distracting or irritating (and often both: distracting at first and irritating after a few slides). Remember that if you use the same effect for every slide, your audience will see it a lot of times. You can fix the time between each bullet appearing as part of the design of your slide, but this is very constraining. If you don't plan to say much about a set of bullet points, it may be better for them to appear together. Sound effects when bullet points appear or when a new slide is presented are widely disliked and can be embarrassing.

3. When you've decided what you want, use the technology well

Let's consider some opportunities. However, as with other parts of this book (for example, when considering word processing, spreadsheets or Web browsing), we do not refer to specific software or describe how to use it.

A lot of people advise you to keep things simple when using presentation software. Well, maybe. Some presentations can appear pretentious, or so highly dependent on technology that genuine face-to-face communication seems to be obscured. In any case, with increasing dependence on technology there is increasing risk that things will not go to plan, perhaps because of a limitation of the system you are using, or because of a system fault or failure. However, the potential benefits are often so great that these risks are well worth taking. Where appropriate you should take opportunities to use the technology to the full. If you are describing a project that has involved developing a piece of software or a website, then you should demonstrate it during your presentation. If you have been working in the lab on an experiment that can only be fully appreciated by watching a video, then show the video clip as part of your presentation. Simpler visual material like photographs, diagrams and drawings should be used whenever they enhance the presentation. Chapter 6 contains advice about diagrams. Sound may enrich the audience experience, even music. But think this through. If music creates a mood at some point, what will happen to the mood when the music stops? If you go for some startling visual or audio effect, you have to be confident that it will work in the context of the presentation. The anticlimax caused by a pause needed to sort out the technology may completely spoil the effect.

Checklist for preparing for a spoken presentation

- Have you thought about your audience: who will be there, and what do they need to know?
- Have you prepared notes (on paper or slides) which will guide you through your content but not prevent you from looking at the audience?
- Have you prepared appropriate visual aids to enhance your presentation?
- Have you worked out what to say about each slide?
- Have you tried out your visual aids?
- Have you practised and timed your whole presentation?

On the day:

- Have you checked the room and the equipment?
- Have you checked that all software you need is available?
- Do you have at least one saved copy of your presentation?
- Do you have a watch?

12.5 Giving the presentation

Basics

Unless the session is very intimate, stand up. Maintain good eye contact with the audience. Don't spend the whole time looking down at your notes, or backwards at the screen, or down to the computer monitor. It doesn't matter who you look at; look at someone, anyone, and keep looking around the audience; try in effect to look at everyone. Of course you will have to look away to operate visual aids or read notes, but then look back.

Don't expect too much from the audience; they won't all be nodding enthusiastically at all your points. This is one of the things you will notice most when you give your presentation – there is not the feedback of normal conversation. Yet you should still look for subtle signs from the audience that your message is or is not getting through.

You must aim to be heard clearly at the back of the room, but don't shout. Try not to speak in a monotone, but don't give your voice unnatural intonation. Most people speak too quickly when they are nervous. If you try consciously to speak a bit more slowly than usual it will give you time to think, and allow you to be heard more clearly. If you need to pause, do so. You can pause for longer than you think without worrying the audience.

In a very large room there may be a microphone available. In some rooms it really is necessary to use one. If it is the type that can be attached to your clothes, it shouldn't cause inconvenience. If it is fixed, it will inevitably restrict your ability to move around. You should not raise your voice when you use a microphone, partly because it is the job of the microphone to make your voice loud enough, and partly to allow you to tune your ear to the sound of your voice coming through the sound system.

Show enthusiasm, try to be yourself, smile when it is natural to do so. A little humour may even be appropriate, but bear two things in mind. You should never *expect* to get a laugh, otherwise you will be put off if you don't. And if you do get one (hopefully when you wanted one) follow it with something serious.

Don't try to be artificially formal or informal. The formality or informality of the occasion will look after itself. You will relax more as the presentation progresses.

Using visual aids

Don't show a beautiful diagram or table (which has taken you hours to prepare) for just a second and then move on to the next slide. You should leave enough time for the audience to understand the content. To leave a silent pause would probably make you feel uncomfortable, so the best idea is to describe clearly what is being shown even if you think it is obvious. The audience are more likely to absorb the information if they see it and hear it at the same time. Think of sports results on the television. They are shown on the screen and read out in full, yet you don't say "Oh shut up, I can read it for myself".

If your diagram is labelled, it may be helpful to design your presentation so that the labels appear one by one, and then you can make sure you say something about each aspect.

If you need to point at the screen, it is possible to use a laser pointer (provided, of course, you have made sure that one will be available). You should practise using one of these first. There is a danger that if the nerves of the occasion are making your hand shake slightly, the effect will be magnified on the screen. The alternative is to point with an arm at the screen. This allows you to make more open gestures, and communicate enthusiasm for what you are showing: "We had to design a special device to go *here*" or "Agreement between the readings is good *here*". It also means that you and the audience are looking at the same thing: it is a shared point of attention. This technique is only suitable in small to medium-sized rooms.

Timing and pacing

This is one of the hardest aspects of all. First of all, it is easy to forget to look at your watch when you start. Also it is easy to say more than you planned, and then realise right at the end that you still have a lot to cover. In that situation the best thing is simply to drop some of your material. Do not try to speak very quickly, or keep saying "I'm sorry, I know I'm running out of time, but I really must describe . . .".

There are no special techniques for time-keeping. Just prepare carefully, practise, and keep an eye on the time.

Style

Dress depends on the occasion. You may feel a little uncomfortable if you are dressed very smartly and nobody else is, but you will feel worse if you are the only person who is not dressed smartly. Feeling smarter than usual will help to put you in the right state of mind. In any case, if you feel you are over-dressed you can do something about it – take off your jacket for example; if you are under-dressed there is nothing you can do.

Don't stand rigidly. Move around and gesture in a natural way. Keep your approach simple and direct. Don't get too concerned about controlling body language or audience psychology. You may need to work harder at gestures and refinements to technique when you have more experience. As a student, if you have something interesting to say, and you have prepared well, your presentation should be a success.

Group presentations

All the material in this chapter is relevant to group presentations. In addition you must make sure that a group presentation is well coordinated. The introduction should name the members and define what each will cover. There should be a prepared handover between group members: "I will now hand over to Ian who will give more detail on the planning of the tests". Use each other's normal familiar names; do not try to be artificially formal by referring to each other as Mr . . . , Miss . . . , etc.

The concluding part should draw together the themes of the group presentation as a whole.

Being videoed

You can learn a lot from seeing a video recording of your presentation.

During the presentation you should try not to be distracted by the fact that you are being videoed. Concentrate on communicating with the audience, not on the camera.

At playback time remember that most people hate hearing their own voice if they are not used to it. Self-criticism is easy; hopefully your lecturer will help you identify your good points.

Answering questions

Responding to a question that you can answer well will be the most relaxed part of your presentation. This is your reward for "knowing your stuff".

If you are asked a question which exposes an area of weakness, don't try to bluff. The best thing is usually to make comments in a related area where you do have some knowledge, but to make it clear that that is what you are doing.

Checklist for self-assessment of a presentation

- Did you look at the audience?
- Did you feel that your presentation interested them?
- Did you communicate your enthusiasm for the subject?
- Did you make good use of your visual material?
- Did you finish on time without rushing?
- Did you say most of what you wanted to say?
- Did you answer questions well?
- What would you do differently next time?

Further reading

Edney, Andrew *PowerPoint 2007 in easy steps*. Computer Step, 2007. Detailed guide with clear layout.

13. Recording experience and personal development

To become a true professional in your field, you must develop the ability to **reflect** – on what you have done, and why. If you apply to become professionally qualified you will have to demonstrate that you have reached the standards set by the professional institution. At work you will be expected to make decisions and judgements about a wide range of subjects; you will be required to assess evidence and may need to justify your decisions later. So keeping a record of what you did, and why, is a must.

Before you reach the stage of recording experience for professional qualification, you may need to keep a logbook as a part of a course of study. This should be treated in the same way as recording experience for professional reasons, and it is not a skill that comes naturally to most people. It needs practice and the advice of more experienced practitioners.

The aim of this chapter is to help you develop skills in recording evidence that will contribute to your studies, to your professional qualification, and to simple good professional practice.

13.1 Logbooks

The first important point about a logbook is that it is not a diary. OK, you need to record the date and time of entries, but it is not a space just to note what happened each day. Another important point is that you should record things as they happen. It's always obvious when a logbook that is being assessed has been written in one go – all in the same ink, with the entries more and more scant as time progresses. That makes the exercise pointless.

This is how a blunt engineer might explain the aim of a logbook. "If you were to step in front of a bus, how would we be able to complete the projects that you were working on?" A logbook is about noting actions, decisions, ideas, calculations, timescales, key sources and contacts, telephone numbers, email addresses and the like; it should also include the reason for recording this information.

The book itself has a particular form too; it must be a bound book – no loose-leaf papers that can become disordered and lost. If possible you should keep your previous logbooks in a safe place, arranged in order.

This may all sound a bit daunting. But you are not trying to record everything you do, just the new stuff, or the things that another person might do differently and so would need to know why you acted the way that you did.

Your logbook should stay with you throughout your working day. It should be to hand when you make a telephone call, when you attend a meeting or when you are solving a problem over coffee with colleagues.

Here are some ideas about the sorts of things that are good to record in a logbook.

When attending a meeting

- Note the people attending – a great way to learn the names of people who may be key to your career. You can even note who sits where at the table; this will help you to recall first names and to appear attentive.
- Note the date and the time of both the start and the end of the meeting.
- Note the key aims of the meeting and any actions that have been allocated to you. Make sure that you also record who expects what and by when, but don't try to record the whole meeting. (There's more about meetings in Chapter 14.)

When making a telephone call

- This applies to interesting and non-standard calls, not a chat to a colleague.
- Record the name of the person who you called (or who called you), their contact details and any actions agreed in the call.

When in the laboratory and workshop

- It can be good to note the environmental conditions; you may need to try to replicate an experiment.
- Note down the details of any equipment used, any people that you worked with and the results that you achieved.
- It is sometimes useful also to record any assumptions that you made; this can be immediately helpful in thinking through the problem that you may be trying to solve. (There's more about recording laboratory work in Chapter 8.)

When on site

- It can be useful to note the state of the site, any significant issues, and progress since your last visit.
- It is important to record any instructions that you give, or are given, and who you spoke with.

When talking about work over coffee

- This can be the most productive of problem-solving sessions, and sometimes happens when you least expect it. If you have your logbook with you, it can be very useful for recording decisions.

When carrying out a piece of research

- Your logbook is valuable for recording your search trail. When you are conducting online research it is amazing how easily you can wander from one subject to another and not realise either where you started or where you intended to end up.
- Record the development of any research instruments, equipment, surveys, questionnaires, data sets and so on.

Figure 13.1 is an extract from an actual logbook by a student working on an industry-related investigatory project.

13.2 Evidence of learning at work and meeting learning outcomes

Some courses involve the assessment of work-based learning. Learning at work is quite different from learning in a traditional academic setting. The classroom presents its own artificial rhythm, but the work situation has real timescales – the need for delivery, and the added pressure of all the other activities that are happening at the same time. In the classroom a tutor will fix a set of learning outcomes, design a learning plan and a delivery timetable, and then set coursework and an examination that test that the learning has taken place. In the workplace, because of the variable nature of the activities and the differences between companies, while the learning outcomes can be fixed, it can be very difficult to define the learning plan.

For example, if you were taking a work-based module in communication skills you might need to meet four learning outcomes like the ones set out below.

1. To be able to demonstrate an ability to use a range of written communication techniques as used in modern business
2. To provide evidence to show the implementation of a range of oral communication techniques
3. To develop a literature review on a topic of relevance to the field of work
4. To critically review your performance in the subject

How would you demonstrate that you had met these learning outcomes? The first could be evidenced through the production of a range of written pieces, perhaps a short report used in the workplace, a logbook, an email communication trail and a more formal letter. Evidence for the second could be the printout of a presentation, testimonials of people who attended, and

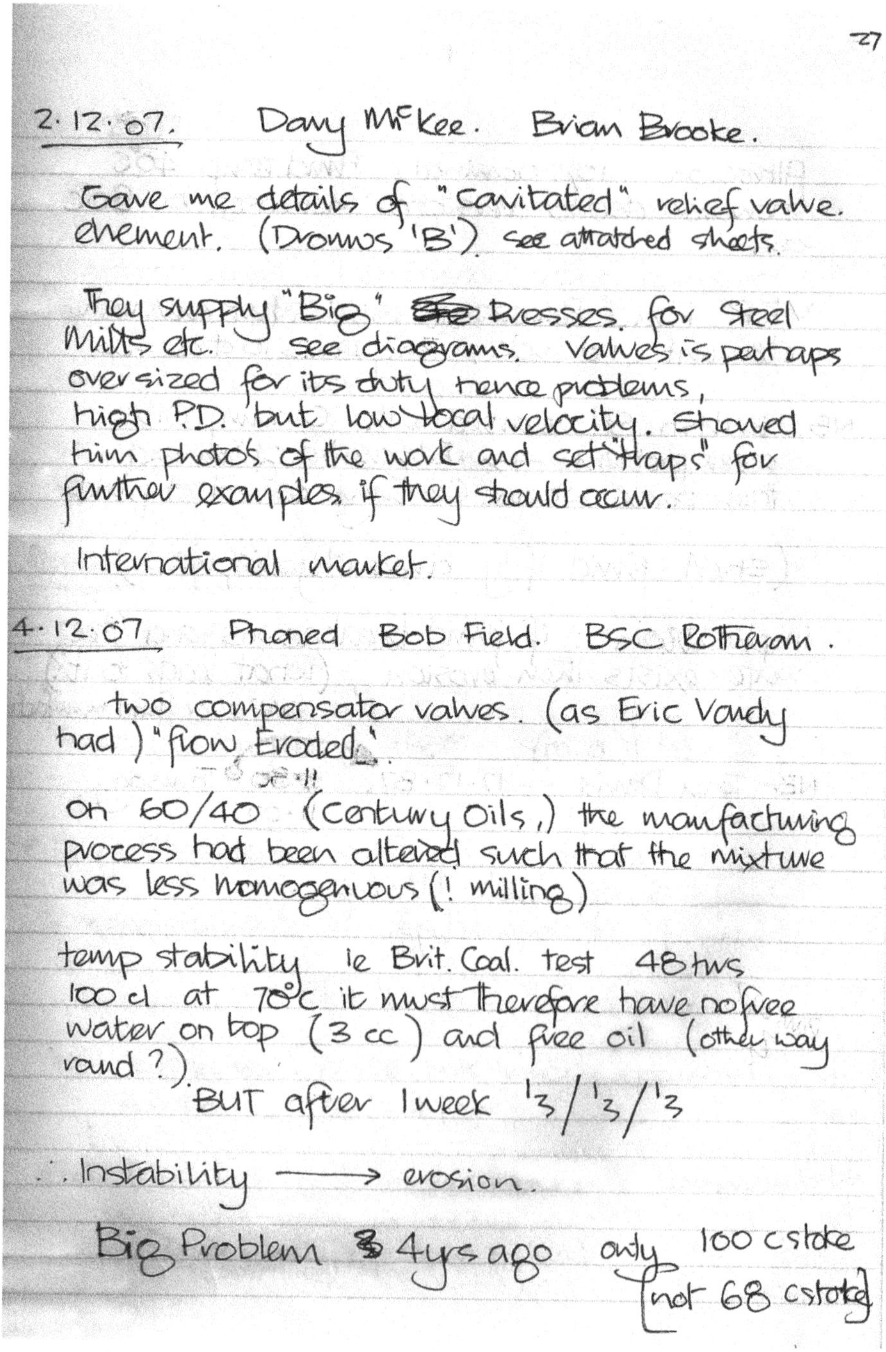

27

2.12.07.   Davy McKee.   Brian Brooke.

Gave me details of "cavitated" relief valve.
element. (Dromos 'B') see attached sheets.

They supply "Big" ~~the~~ Presses. for Steel
Mills etc.   see diagrams.   Valves is perhaps
oversized for its duty hence problems,
high P.D. but low local velocity. Showed
him photo's of the work and set "traps" for
further examples if they should occur.

International market.

4.12.07   Phoned Bob Field. BSC Rotheram.

   two compensator valves. (as Eric Vandy
had) "flow Eroded".

On 60/40 (Century Oils,) the manufacturing
process had been altered such that the mixture
was less homogenous (! milling)

temp stability ie Brit. Coal. test   48 hrs
100 cl at 70°C it must therefore have no free
water on top (3 cc) and free oil (other way
round?).
      BUT after 1 week $\frac{1}{3} / \frac{1}{3} / \frac{1}{3}$

∴ Instability ⟶ erosion.

   Big Problem $ 4yrs ago only 100 cstoke
                              [not 68 cstoke]

Figure 13.1 Extract from a student logbook
Source: © Malcolm Blake

perhaps even a video. The third is more academic in nature and would be based on a piece of research. This should be related to a live project in the workplace and with real practical value. The fourth outcome would be demonstrated by an evaluation of the skills you have developed; this might be in writing, but could be an audio recording. It could equally be the views of others, perhaps managers and colleagues, presented in such a way as to show your ability to critically review your own performance.

When demonstrating learning in the workplace you should remember that evidence of learning is rarely exactly the same as simply presenting the outcome of the work itself. You will normally be required to review, or reflect on, what you did and why. The reason for this is that academic learning is about understanding why things happen, not just how things are done. It is quite normal to learn how to do something and yet to have little idea about why it is effective. A good student wants to know why an approach is effective, and will use this learning to improve performance the next time a similar problem is faced. This is an essential element in good problem solving.

13.3 Personal development planning (PDP)

Throughout this chapter there has been an emphasis on **reflection** and obviously in the process of personal development planning (PDP) reflection is key. Where PDP is part of an academic course, especially when it is a requirement or when it is assessed, there is a need for evidence. This is likely to include evidence of the reflection itself (which as we have seen already in this chapter is a hard thing to produce); evidence that the reflection has led to planning, which in turn involves defining needs and goals; and evidence of learning and achievement.

People will talk to you extensively about ensuring that you maintain a portfolio of your work together with your thoughts about that work, in order to plan your personal development. The thinking is quite simple: if you record what you do and why, then you are able to learn and develop more quickly. Another benefit is that you can identify gaps in your learning and so plan your personal development.

This links in closely with maintaining a logbook, where you record activity and contacts etc. The portfolio, however, can contain examples of your work, good and bad, especially if they show how you have learned from the bad and developed into the good. The portfolio can be paper-based or electronic, and is yours. It may follow a very tightly prescribed format, or may be a compilation of various types of material selected by you. The focus can be on academic, personal and/or professional development. There is plenty of guidance around and we include some in our **Further reading** suggestions. In the end, the portfolio and the learning you gain are only as good as the effort you put in.

As with a logbook, the skills developed are just as relevant to the work of a professional as they are to that of a student. Most professional people are required to keep records of their personal development, especially continuing professional development (CPD). In this case reflection is again identified as key. The worst sort of CPD record is a list of courses attended, and the best is a living reflection on learning and future needs.

So there is no standard or typical personal development portfolio, and the best approach is to follow the advice you are given and to get ideas from your reading.

Further reading

Cottrell, Stella *Skills for success – the personal development planning handbook.* Palgrave, 2003. A real self-help book showing how personal development planning can lead to personal benefits and success.

Stefani, Lorraine, Robin Mason and Chris Pegler *The educational potential of e-portfolios – supporting personal development and reflective learning.* Routledge, 2007.

Higher Education Academy website sources on PDP.

14. Group work and meetings

14.1 Problem solving and group work

In most applied situations we solve problems in an hourglass-shaped way:

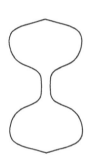

Problem or market need

Creative thinking, information gathering and lots of ideas

Filtering the good from the bad and the plain mad

Development of the best ideas, more information gathering and creativity

Final choice of the working solution; development and communication

activities in the wide parts of the hourglass need open thinking, and activities in the narrow parts need focus.

At each stage, while we may be working in the same team, we need different combinations of skills in order to ensure that we gain the maximum advantage from the activity. We need people sharing ideas and using their natural curiosity during the information gathering phases; these are the sections where the hourglass is at its widest. During the final stage we need to be sure that we are using each person's special knowledge and skills to the best advantage and this needs to be coordinated and project-managed.

This leads us to think about how groups work together and about the roles that people take in a team. There are inventories and tests that will guide you to understand how you work best in a team. It is a very good idea to take one of these tests. It will give you some useful insights into how you perform best with other people. Some of us are best at organising, others at idea generation, others at doing the background work, and each of these activities is essential for any team to operate.

At times you will be working with your friends in a group during your study; this may seem like a good idea because you all get on well together. But working with people who are like you may detract from the quality of your group work. You may each have to take on roles that are not naturally yours, and this can cause you to stretch your friendship. Sometimes it can be a good idea to work with people that you don't see as naturally part of your group but that you have noticed to be good at getting their coursework in on time, or that are great at solving complex mathematical problems or that are natural communicators and are able to put ideas across well.

We need a whole range of skills to solve problems and to work in a team; and we need to think about the type of communication that we use as we move through the life of a project.

14.2 Communication as part of working with people

One of the factors that determines the quality of a group's performance is communication. A sign that a group is communicating well is that it is good at making group decisions and also good at dividing workload between members. There are suggestions in the next section about how to manage decision making, but first, here are some general comments about communicating in groups.

Listening

When working in a group, an important communication skill is listening; after all, during discussions in a group of four, you are likely to be listening three times longer than you are speaking. As a colleague puts it, you've got two ears and one mouth. Listening well involves respect for other people's ideas, willingness to encourage them to make their contributions, and concentration. Listening, in a different context, is also an important study skill for students.

In industry, technical staff often work in multi-disciplinary groups. This makes particular demands of their communication skills in every respect, as they may not always be good at communicating with people outside their discipline (and remember, good communicators are not just good at giving out messages but also good at receiving them). In engineering, for example, multi-disciplinary groups are often led (to many engineers' disappointment) by non-engineers. A quantity surveyor, who was leading a project team which included engineers, was asked what he thought he was better at than the engineers. He said, "Listening".

Assertiveness

Working with other people cannot always be based purely on friendship and cooperation. It is often important in working relationships to know how to be assertive. You are assertive when you seek mutual respect. The respect you show others is as important as the respect you expect from them. You should not be aggressive or bossy, nor should you be passive or defensive.

Informal communication

Much of this book is devoted to reports and spoken presentations, but they are not the only means of communication for students and professionals, especially when they work in groups. A great deal of information is communicated informally: in conversations, as scribbles on scraps of paper, or on the phone. We should not try to apply the same principles to informal communication as we do to formal communication. We would have a strange conversation if we avoided using all unnecessary words. If colleagues say "Good morning, how are you?", and you reply "Please use only necessary words", they will think you are pretending to be a robot.

Much communication does not involve words at all. Gestures, posture, facial expression and tone of voice can communicate a great deal. Some communication does not aim to get across a message but to establish a good working relationship or inspire confidence.

14.3 Making group work successful

The key to making group work successful is that *you* can influence your group – not necessarily by being bossy or showing "leadership", but by making good suggestions. A group that is not working well wastes *your* time, so it is worth doing your best to influence how it operates in a positive way. You should never give up and let things happen, but you don't need to fear that the only way you can influence what happens is by "taking control".

As an example, if someone else is being over-dominant, it may be best to find a way of getting the group as a whole to sort this out, rather than trying to "take on" the individual yourself. In other cases, if the discussion becomes intense – even heated – this may simply be a healthy sign that the group is becoming fully involved in its task.

It also helps to be aware of the fact that everyone behaves differently in groups, as we have discussed. It is important for the group as a whole to make best use of the varied talents it contains, by, for example, encouraging quiet members to contribute, while not stifling people with ideas. Don't think a "listener" isn't contributing; if someone is *not* listening they are not helping at all.

Certainly self-awareness, by all group members, is important. Ideally a group should contain a balanced range of characteristics. A natural leader knows what motivates people and how to bring every member's skills into play. Allowing a person who behaves in a dominant manner to "lead" may seem initially attractive but can cause stifling of creativity, non-participation and a transfer of responsibility which negatively affects performance.

One way you can make a positive influence is by proposing principles (early on in the life of the group) by which the group should conduct its discussions, or manage its decision making. You could say, "Here is a list of ideas for group work that I've found in a book", and then show them the following list.

1. Have your discussions at properly planned **meetings**.
2. Before the discussion starts, agree the topics you wish to discuss and your aims in discussing them (an **agenda**).
3. Also before the meeting, agree who should takes notes (**minutes**). You could also decide that a particular person will **chair** the meeting (though that may be too structured for many situations).
4. As the meeting proceeds record the decisions and **actions** (what will be done, who will do it, and by when). Agree clear expectations of each member's contribution.

Chairing and taking minutes (in formal meetings) are discussed in the next section.

If your group can agree to conduct its discussions in this structured way, most problems of coordination and efficient working will become much easier to solve. At each meeting you will agree what must be done and who will do it. This may not be easy, but it will be clear to everyone that progress cannot be made until agreement has been reached. Obviously the split in workload must be fair, and ideally each job should be given to the person most able to do it. At the subsequent meeting each member should report back on progress.

It is important that decisions are recorded – otherwise plans and deadlines will drift. Another benefit of good records is that they impress your lecturers. Even if they are not required or expected, submit the records of your meetings to show how well your group has operated.

One device for improving the operation of your group is to make one of the discussion points at a meeting: "How well are we working as a group? Could we improve the way we work together?" If a group is working well already, this sort of discussion is likely to make things even better. However, if a group is not working well you must be careful not to allow this sort of discussion to become so confrontational that it is counter-productive.

14.4 Meetings

The points in this section are valid for any sort of meeting, but will relate mainly to more structured ways of doing things.

The agenda

If five students are working together on a group project, an effective but still quite informal agenda might look like this.

Agenda

Meeting, 23 November, 2 pm, Project Room

1. Update on progress: Toni, Gabba, Sukhi, Ali, Sam
2. Costings on pre-design ideas
3. Are we ready to start detailed design?
4. Detailed plan for the next two weeks

The agenda for a more formal type of meeting is likely to look like this.

Agenda

Course Committee Meeting

5 March, 1 pm, Committee Room 3

1. Apologies for absence
2. Minutes of the last meeting, 1 February
3. Matters arising
4. Issues raised by students
5. Issues raised by staff
6. Discussion of new education proposals by the Engineering Council
7. Progress on the new library building
8. Any other business
9. Date of next meeting

Chairing

Appointing someone to chair a small meeting about group project work might seem too formal to many students. But there may be circumstances in which you decide it would be helpful, or you may be required as part of the project brief to hold properly chaired meetings. In larger meetings it is hard to make progress without having someone to chair. Here are some brief notes on this important job.

Preparing

This depends on the circumstances, but preparation may include making sure that the time and location have been fixed, and that participants have been given the papers in time (for a formal meeting, usually the agenda, the minutes of the previous meeting, and other supporting papers).

Planning the discussion

Imagine that an item for discussion is "Plan the campaign for attracting sponsorship". If you (the person chairing) have given this no thought in advance, you will probably just say, "OK, on to the next item, planning the campaign for attracting sponsorship. Any ideas?" The discussion that follows will be unstructured, in random order, and very difficult to control or draw towards useful conclusions. If you think about it in advance, you may see a logical order for the discussion, and be able to say, "OK, on to the next item, planning the campaign for attracting sponsorship. I suggest we discuss this important task in the following order: (1) approaches that have been successful in previous years, (2) organisations to approach, and (3) timetable. Is that OK with everyone?" If someone suggests another good topic, you can include that in the plan too.

Guiding the discussion

You must try to ensure that the discussion has quality and focus, that time is used efficiently, that consensus is identified, and that the outcome has value. When a decision is reached, you must be satisfied that it has the support of the meeting, and that it has been recorded.

A very important skill is being able to curtail over-lengthy contributions (in other words, tell people to shut up) without causing offence. You should have no qualms about doing this. If you think someone has been going on too long, it is likely that everyone else at the meeting (except the speaker) thinks so too. Some people don't know how to round off their contribution, so once you've understood what they're trying to say, interrupt them with your own positive-sounding summary of their point, "Yes, thanks Graham, that's very helpful – we must make it clear to local companies that they could gain financially from their involvement in the project".

Participating

Effective participation in meetings needs work too. Before the meeting, you should read the papers carefully and think about any points you wish to make. If you are a representative, you should think about whether you need to consult with the people you represent – before or after the meeting. If your contributions to the meeting are brief and to the point, people will pay more attention to what you say than if you ramble. Obviously a large meeting can only be effective if one person speaks at a time. In completely formal meetings participants are given permission to speak by the person chairing, and they address their contributions to that person.

Minute-taking

Minutes are not word-for-word (or minute-by-minute) records of a meeting. They are summaries of important issues, of decisions, and of what individual people must do after the meeting (**actions**). The record of an informal meeting of students working on a group project (like the one on 23 November, agenda above) might look like this.

Project meeting – 23 November, 2 pm

Toni, Gabba, Sukhi, Sam (Ali not there: at job interview)

1. Update on progress

   All targets set at last meeting OK – costings now complete.

2. Costings

   Decided these are OK, but must combine on one spreadsheet.
   Action: Sam, by 30 Nov

3. Start detailed design?

   Agreed – feasibility stage complete. Costings show Option B is best choice.

   Agreed – start detailed design of Option B. Must complete feasibility report, with costings included.
   Action: Suhki, by 7 Dec (assuming spreadsheet received by 30 Nov)

4. Plan for next two weeks

   Toni: preliminary structural calculations

   Gabba: modelling of tracker mechanism

   Ali: legal aspects (contact law lecturer who came to briefing meeting)

   Sam: update detailed work plan

Next meeting, 7 December, 4 pm, Project Room – important, last before break (Ali buys drinks after)

More detail may be required for a larger formal meeting. But taking minutes does not require speed-writing, it just involves listening carefully, and keeping track of issues and decisions. A record of those present should be made (commonly by passing round a sheet for people to sign). A section of the minutes for the Course Committee Meeting on 5 March (agenda above) might look like this.

Minutes

Course Committee Meeting
5 March, 1 pm, Committee Room 3

Present:

Staff – Dr John Roland (Chair), Dr Mary Wiggins, Dr Peter Ubiaro,
    Prof Paul Castle
Students – 1st year: Suhki Sidhu, Misha Stander
        2nd year: Jim Holland (Minute Secretary), Ali Hall
        3rd year: Ruth Pearse

1. Apologies for absence
from Prof Prosser, Jacob Chat (3rd year)

2. Minutes of the last meeting, 1 February
These were accepted as an accurate record.

3. Matters arising
Dr Roland reported that the ventilation system in the computer suite had
been repaired. He had still not heard from Estates Dept about late access
to the studio, and would chase it up again.   **Action: Dr Roland**

4. Issues raised by students
.1 The 2nd year students commented that the input from outside experts
at the start of the design project had been helpful, but that it would be
very useful if the same experts could attend another session, before the
end of the project, to comment on the designs being developed. It was
agreed that this was a good idea, and Dr Ubiaro said he would arrange it.
                                                    Action: Dr Ubiaro

.2 The 1st year students asked why the management coursework
assignment . . .

Copies of the minutes of a meeting should be distributed to all members,
including those who were absent.

Further reading

Back, Ken and Kate Back *Assertiveness at work: a practical guide to handling
awkward situations*, 3rd edition. McGraw-Hill, 2005. A comprehensive and
helpful book.

15. Essays and exam answers

You will not write as many essays as an arts student, but you will probably write some. As well as coursework essays, exam questions in management or economics may call for short essays. Science subjects like materials science or geology will involve descriptive exam answers that have much in common with essays. After you graduate you may take exams that involve writing essays in order to gain a professional qualification.

In this chapter we will deal first with essays generally, then with the special problems of writing in exams and exam technique in general.

15.1 Essays

The process of writing an essay

No one writes essays at work. Professional people and students in technical areas come across essays only as forms of learning and assessment. The essays they write are descriptive rather than mathematical, and usually include an element of discussion. Essays are personal, not simply factual, so while the assessor may look for a number of specific points, there will be no single correct answer. However, while your essay may include your opinions, these must be based on facts and not on flights of fancy. Essays call for knowledge, careful and balanced interpretation, breadth of thought, and good writing technique.

When writing an essay you must take care to write clear English, using the ideas of Chapters 2, 3 and 4. You should keep the style formal, and the sentences generally short (unless you feel confident writing in this format).

The stages of the writing process described in 5.1 (page 36) apply perfectly well to essay writing. The main activities when you prepare, *defining the task* and *sorting out ideas*, revolve around the title of the essay. That is a characteristic of essays; while the aims and scope may not be defined for you, the title usually is. As you develop ideas, you are really opening up the title, and, as with all writing, this is where the scope and richness of your content are determined. We will look at an example later.

The development of ideas is linked to researching the topic. It is not possible to generalise about this. If the topic requires you to evaluate arguments that you understand well already, you may want to develop your ideas to an advanced stage before researching particular examples or case studies. If your factual knowledge is insufficient to allow you to develop ideas, you will need to do some reading first.

Structuring your ideas involves establishing the themes of your essay. This may involve grouping points for and against a proposal. The result will be a plan for your essay. The basic unit in essay-planning is usually the paragraph. You should plan your essay paragraph by paragraph.

The best ways to *write* and *improve* an essay are as described in 5.1.

You must be particularly careful about the way you use material from other sources. If you quote facts, you must say where they came from. Text that is quoted in full must be included in inverted commas, and all references to other people's work must be acknowledged carefully. (More information on references and quotations is given in 7.2, page 66.)

Essay structure

Standard advice is that there should be three sections: introduction, middle and conclusions. "Middle" doesn't help much, it really just means the main content of the essay; but the other two divisions are worth considering here.

The first paragraph (or more) should be an introduction to the topic and to your treatment of it. If you have chosen to interpret the title in a particular way, or to concentrate on one aspect, you should explain this at the start.

In your last paragraph (or more) you must give some outcome. Don't overdo it by writing something like: "I conclude that all computer scientists should learn Japanese and own helicopters". Just try to resolve the arguments in a realistic and helpful way. Make the reader feel that something worthwhile has been achieved.

Opening up the title

This is the key to worthwhile and imaginative essay-writing. Let's take an example that requires no specialist knowledge beyond the normal experiences of a student. Suppose you have been asked to write an essay with the title **Technical courses are too narrow**. The title, of course, is a controversial assertion, like the motion in a debate. You are expected to consider the arguments for and against – in effect to answer the question "Are technical courses too narrow?"

It's time for ideas, time to open up the title. Write the title at the top of the sheet of paper and start thinking. Write down all your thoughts in note form. Here are the thoughts you might have (written out in full).

Who decides how narrow courses should be? The answer must be: the lecturers who teach them, the profession (which approves courses via the professional institutions), and the government (which ultimately controls the length of courses).

What is the aim of your course? Is it just to prepare you for industry, or is it to provide an education for life? Society tends to expect professional people to be "educated" as well as simply good at their jobs. In any case not all graduates get jobs in their own discipline.

What subjects can widen your type of course? Economics, Management and Communication seem to be common. Some courses have the option of a foreign language. Wider? How about Business Studies, Psychology or Sociology? Wider still? Politics, Philosophy, History, Art, Literature?

But courses seem to be crammed full of science and applications. There is scarcely time to *think*, let alone study "wider subjects". Could some of the technical content be dropped? If so, what? Perhaps courses need to be longer.

Technical people seem to be thought of as having narrow interests and being reluctant to become involved in public life outside their profession. This may be why they lack influence and political clout. Is this connected with the narrowness of their education?

Yet can the experience of any student (meeting people, playing sports) be described as narrow?

An alternative to writing notes down the page is to set out your ideas on a "splay diagram". This allows you to develop ideas and to show the relationship between them. The ideas we have considered about our essay title could be represented by the splay diagram in Figure 15.1. Many people are enthusiastic about these diagrams and use them not only for essays but also for exam answers, revision, lectures, projects, and any other activity in which ideas are being developed.

We've opened up our essay title to cover quite a wide range of interesting points. It's probably time for some researching. How different is one course from another? Some courses have different lengths; how is the time used? What about different countries? Have the professional institutions or the government published their views? What do your lecturers think? Off to work – there's obviously no shortage of ideas.

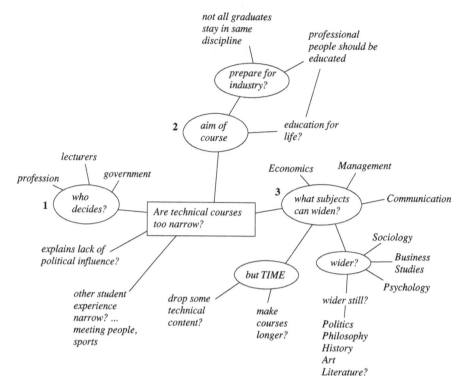

Figure 15.1 Splay diagram

15.2 Writing in exams

Some exams in technical subjects call for answers in the form of essays or extended descriptions. Students in those subjects don't get much practice at descriptive writing in exams, and often don't do it very well. Here are some ideas that may help.

First, think about the question. Watch out for optical illusions and self-deception. Think about the question that is actually set, not the one you hoped for, or the one you would be able to answer easily. One of the things the examiner will be looking for most is how closely you answer the question. If you don't understand the question, don't answer it (assuming there is a choice, and you do understand some of the others).

Then you should begin a condensed version of "the writing process". Because of time constraints, a free-flowing ideas session is not really possible. This means you must work out the general themes or topics before you think of detailed ideas. Then as you note down the details they can form part of a ready-made structure. But *make notes* – that is the important thing. When you have completed this stage, write out your answer. Keep your style formal and your sentences short. Don't try to impress. Be factual and specific. Use diagrams and examples wherever they may help to make your meaning clear. Don't try to make jokes – the only thing that will cheer up the examiner is a good answer. When you have finished, no matter how short of time you may be, check what you have written thoroughly.

Exam technique

Here is some brief advice on exam technique in general. It applies to any type of exam.

Plan your revision

Work out a plan for your revision long before the exams. Make notes or draw a chart to remind you of what you have decided. Include sufficient time off revising, so you don't wear yourself out and become less effective at revising. Adjust your plan when necessary to keep it up to date.

Look at past papers

Past papers can certainly help you prepare, but check that there will be no major changes in subject content or structure from previous years. If you have been told there will be a significant change, ask for samples of the new format.

Be prepared

Don't let something trivial ruin your composure just before the exam. That means you should:

- double-check the time and location of the exam
- be sure you have everything you need (including spare pens and calculator batteries)
- plan for possible travel delays.

Don't overdo it just before the exam

Like an athlete preparing for an important event, do some gentle work the evening before and then relax (this means that your revision must have been well planned). Don't join the overexcited group of fellow students outside the exam room asking each other questions like "What's the definition of . . .", "How do you work out . . . ?" – you will only waste mental energy and make yourself anxious.

Read the paper carefully at the start of the exam

If there is a choice of questions, choose wisely. Start with the questions you think are easiest.

Always be aware of time during the exam

Plan your time, and monitor your progress, carefully. Allow a generous amount of time for reading the paper at the start and for checking at the end. If the mark allocation for a question is given, use it to proportion your time. Don't allow much over-run for a question – if there's time at the end you can go back.

Be careful with calculations

In questions involving calculations (the norm in many engineering and applied science subjects) use every opportunity to cross-check answers and confirm that calculated values are realistic. Always give units. Set up calculations so they can be followed easily (then you can get marks for the correct method even if you make a numerical mistake).

Further reading

Northedge, Andrew *The good study guide*. Open University, 2005.

16. Letters and email

A lot of relatively formal communication takes place by email, but letters still have a place. Students certainly need to demonstrate letter-writing skills in covering letters for job applications, and these are covered in Chapter 17. More general aspects of letter writing are covered here, followed by the approaches needed when writing formal emails, and by some general advice about everyday use of email and social networking.

16.1 Letters

Format

There are well-established conventions for the way a letter is set out. If you do not follow them you will risk losing your reader's confidence. General appearance is also important; an untidy letter may not be taken seriously.

Angela Fulchrum is a final year student on the BEng in Mechanical and Manufacturing Engineering at the University of Devon. She is carrying out a project on the modelling of manufacturing efficiency. At a careers fair recently an employer encouraged her to contact someone in his organisation (a medium-sized manufacturing enterprise) to arrange a visit to gain some first-hand knowledge. "I suggest you write a letter in the first instance to Derek Greene, our manufacturing manager", he said. Here is her letter.

18 Bere Crescent
Souweston
Devon SU8 2ER

25 January 2011

Mr D. Greene
Manufacturing Manager
Medium Makers
Halo Industrial Estate
Burridge
Devon BU7 4RT

Dear Mr Greene

Final year student project

I am a final year student on the BEng in Mechanical and Manufacturing Engineering at the University of Devon. I am carrying out a project on aspects of the modelling of manufacturing processes. I am keen to visit local manufacturing plants in order to increase my understanding. I met Andrew Davison at the South West Careers Forum recently and he advised me to contact you.

I would be grateful if you would allow me to visit your plant. I appreciate that you would want such a visit to be brief, and I would try to ensure that I did not take up more than one hour of your time. The most suitable time for me to make a visit would be during this term: between now and 2 April.

If you are able to agree to this request, please could you suggest a time and date.

I will not make any reference to your plant in my report without your permission. I would be happy to give you a copy, if you wished to have one, when the project is complete.

Yours sincerely

Angela Fulchrum

Angela Fulchrum

A.Fulchrum@devon.ac.uk

Let us consider the format of letters in detail, with reference to the example above.

Your address

It is normal to have no punctuation in an address. Crescent *could* be abbreviated to Cres, Road to Rd, but why? To save less than a second of your time? If you are typing your address, it usually goes in the top right of the page. If you are using headed notepaper, you will not need to type your address. It is helpful to give your email address; a suitable place is below your name at the bottom of the letter.

Date

If you have typed your address at the top right, the date usually goes underneath it. With headed paper, the date can be written on either side (it might depend on the design of the stationery). The best format is: 11 February 2012. There is no need to write 11th. The month and year written out in full looks better than 11 Feb 12. The American convention is to write February 11, 2012. That is a good reason for not writing 11.2.12, which could be understood by an American to be November 2.

Their name and address

You would not include the recipient's name and address in an informal letter (except on the envelope). But it is important to include it on a formal letter so that you have a record, on the copy that you keep, of where the letter was sent. Also, formal letters are often sent in envelopes with windows, with the notepaper folded so that this name and address is visible when the envelope is sealed (and does not have to be printed again on the outside).

The normal position for the recipient's name and address on the notepaper is on the left, above the "Dear . . .".

Dear . . .

Angela knows the name of the person she is writing to. If she had not known the name she would have had to start her letter "Dear Sir/Madam".

If Derek Greene replied by letter he would know he was writing to someone called Angela Fulchrum. He would not know if it was appropriate to write "Dear Mrs Fulchrum" or "Dear Miss Fulchrum". "Dear Angela" would be fine in this situation but maybe not if the letter needed to be more formal. He could use the safe, fairly formal "Dear Ms Fulchrum". Or he might prefer "Dear Angela Fulchrum". This is a useful alternative, and is particularly helpful when someone has given only surname and initials or has written their first name as Chris or Nicky (and therefore could be male or female), or has a name whose form is unfamiliar to you.

If you are replying to someone whose name is printed below the signature on their letter as, say, E. Stuart (Mrs) or Dr J. Robinson, you should write "Dear Mrs Stuart" or "Dear Dr Robinson".

Heading

A heading (like **Final year student project**) helps to clarify the purpose of the letter. The heading is written below the "Dear . . .", and should be bold or underlined.

Opening

If your letter refers to previous correspondence you should make this clear. After this, you should come to the point of your letter quickly.

Closing

There is no need to "round off" a letter with a special closing remark.

Yours . . .

Letters that start "Dear Sir/Madam" end "Yours faithfully". Letters that are addressed to someone by name end "Yours sincerely".

Style and length

The language in letters should be clear and concise. There is no special "letter language", or "business language"; you should never *try* to be formal or you will become pompous. Nor should you be over-polite or obsequious. If you say that you "would be very grateful" too many times in a letter, you will sound insincere. However, you should be careful not to use a style which is so informal that it sounds casual.

You should make your letters no longer than is necessary. Most letters fit on one page. If you have a great deal of information to communicate, it is usually better to present the information in a self-contained note or report, and write a covering letter which explains what the note or report is about and why it is being sent.

Word processors offer standard "styles" (formats) for letters.

Checklist for a letter

Have you included:

- your address
- the date
- their name and address
- a heading
- their reference (if appropriate)?

Have you followed the convention:

- Dear (name) – Yours sincerely
- Dear Sir/Madam – Yours faithfully?

Does your letter come to the point quickly?
Is it concise?

16.2 Email

"Serious" emails

You will be used to communicating by email with friends and lecturers. Electronic communication is the norm in the industrial environment too, and it is likely that while you are a student you will need to contact someone in industry by email – maybe about a project, a "mentoring" scheme, or a job vacancy. So should you use a more formal, "serious" style than you do with your friends? The answer is: yes, of course! The style of any communication – phone call, phone message, letter, note – varies according to the person you are communicating with.

In spite of that, email is not as formal as letter-writing, and there are fewer well-established conventions of layout and tone.

Let's imagine that Simon Benton is a Civil Engineering student carrying out a final year project on river weir design. Dr Anthony Lewis is a researcher whose interests (as proclaimed on his Web page) include flow past bridge piers. Simon wants to find out if any of the work might be relevant to weir design.

There may be fewer formal conventions in email, but Simon sensibly guesses that "Hi Tony" might be an over-familiar opening. The content of his message is:

Dear Dr Lewis

I have seen from your Web page that you have carried out research on flow past bridge piers. I am a final year Civil Engineering student at the University of Devon and I am carrying out an investigatory project on river weir design, including the impact of pier structures on the river flow. Could you tell me if your work has included use of the methods of analysis by Yarnell or Rehbock? I would also be very grateful if you could recommend any publications, by you or others, that refer to work in this area.

Best wishes
Simon Benton

The "Dear . . ." is not a strict convention. If first names were being used, the message might simply begin "Anthony" rather than "Dear Anthony". But there is a general feeling that "Dear Dr Lewis" has a politer feel than just "Dr Lewis".

It is not normal to end "Yours sincerely" as in a letter. "Regards", "Best regards" or "Best wishes" seem to be the most common endings.

The message is followed by Simon's "signature" which includes all his contact details. At the start of the message, Simon enters as the "Subject": **Enquiry about bridge pier research**. An informative subject is a help, because when people are busy at work they may put off opening messages that do not have clear subjects.

Generally the content of Simon's message is brief and to the point. Email is a convenient form for answering specific queries because the replier can copy the questions from the original message and type in the responses. You are probably more likely to get an answer to this sort of query if you make it by email than by letter. (Of course, if you make a poorly thought-out "please could you help me with my project" type of request, you are unlikely to get a helpful reply to either an email or a letter.) Angela wrote a letter to Derek Greene (at the start of this chapter) because that had been specifically suggested to her.

Take care

1. We all like the speed of email, but it creates some dangers. You have to discipline yourself to check an important email carefully before you send it. There is something about writing letters that puts us in the mood to be fussy about details. We are more likely to send emails quickly. But emails can have formal functions, including the important types of request we have already discussed, and in those circumstances accuracy and correctness become very important. If you are making a job application, it is just as important to

avoid a spelling mistake in an email as it is in a letter. Most email packages have integral spell-checkers, but remember also that you can copy and paste into email software, so you could word-process important messages before sending them.

2. We like the informality of email, and it is sometimes difficult to adjust to a more formal approach with someone we don't know. But it is fairly obvious that too much informality would make Dr Lewis less inclined to give a serious answer about bridge piers.

3. Sending attachments can be very useful, of course. Remember that the recipient must have compatible software to use the attachment, and that you must check for viruses before sending. Because of the fear of viruses, some people who receive a lot of unsolicited emails (in an HR department, for example) set up their system to reject messages with attachments from people who are contacting them for the first time. If you wish to email your CV as an attachment, it might be best to make contact with the company first using a message with no attachment.

Everyday emails

We have considered the fairly formal use of email, and in those circumstances we know we must try to be careful about what we do. But email is very much part of everyday life for students and professionals. It is quick and easy – indeed many of us now email from our mobile phones and so don't even need a computer. The ease with which we use email means that it is easy to make some very simple but potentially highly damaging mistakes. Here are some recommended ways of doing things (or not doing things), as a student and as a professional, that may reduce these dangers.

You should always stop yourself from responding immediately to an email that has made you angry. A passionate response to a crass email will always lead to greater heartache for both parties. Think to yourself that the other person may not have meant to say the things in the way that you have read them, and that if you were to meet face-to-face you would both be able to use body language to tell a great deal of the story.

Another dangerous mistake is to hit the reply-all button when you intended only to respond to one person. Now everyone knows what you think of a colleague, including that colleague! In fact, because any email can be forwarded to anyone, you should never assume that comments in an email are confidential.

The most important thing about email communication is that it should be short. There is no reason why you should expect a whole group of colleagues to read your ramblings. The fact that email is quick and easy does not give us the right to dump our thoughts on other people. A few sentences are plenty. If you have more to say, then an attachment is normally much easier to read. If there is no alternative, then apologise for the length at the beginning of the email.

Social networking

The advent of social networking sites has had a major impact on the way we communicate. Who would have thought that to limit yourself to 140 characters in a tweet would have become such a craze! To be able to post all of those amazing holiday photos online for all of your friends to share is great, but don't forget that you are also making them available for others to see. Those others may be your employer, the family of your boyfriend or girlfriend, or your academic tutors. If you had a great time, it's fine to let everyone know – as long as the photos don't compromise you or others.

At work the risks of careless communication can be great. The comment you post about your company, which was meant as a joke between mates, may lead to a drop in sales – that will not be funny if it costs you your job.

Social networks are amazingly powerful tools; you can use them to advertise your skills and to get a good job. Personal development planning (PDP, see Chapter 13) is important, but equally important is how you then use the output; social networks can help in this. But you need to be highly aware of the importance of maintaining a distinction between how you use social networks in your private life and how you use them to promote your professional life. Remember how people in the serious glare of publicity can be brought crashing to earth for doing quite unexceptional things. Don't forget we all have a capacity to judge by different standards when we observe others!

17. CVs and job applications

CVs and job applications are important tests of communication skills; that is why they are considered in this book. This chapter is designed to help you write CVs and job applications, but it is not meant to be a comprehensive guide to finding a job. For the full story you should listen to the guidance that is given by the careers advisers at your institution.

They will explain that there are several essential stages before a job application becomes a writing task, including identifying your main skills, analysing the job description, finding out about the employer, and thinking about which of your attributes are likely to be of most use in that particular job.

Your application – its appearance, and the care, thought and imagination that have gone into its preparation – can really help you get an interview. When there are many applicants with similar qualifications, the impression given by the application may be all an employer has to base decisions upon. You should always try to see your application from the employer's point of view.

You will want to make the most of what you have, and to present it in the best light. If you think there is something special about you, you must make sure that it is obvious in your application. But do not try to make something that is not remarkable sound as if it is, as in the following extract from a student's draft application. "Having studied at university for three years I believe the knowledge I have gained would be a very considerable asset to your organisation." In referring in this way to his degree – a qualification which all applicants for a graduate vacancy would have possessed – this applicant, named Jeff (more about him later), was trying too hard. Yet Jeff had given no prominence to the highly relevant fact that he had run his own business for two years before going to university.

A CV is normally used for a job application when an application form is not mentioned in the job advertisement (the wording might be "applications should be made in writing"), or when your application is not in response to a specific advertisement. CVs can also be asked for at the end of the online application process. The contents of a CV and an application form are similar. Let's consider the CV first as it is a greater test of communication ability. Many of the points will also be relevant to application forms.

Whatever format your application takes, keep a copy of what you send for your own records, and to help you prepare for an interview.

CVs

CV is short for *curriculum vitae*, the Latin for "the course of life". The word *résumé* is generally used in North America. There are a few alternative formats.

We met Angela Fulchrum, final year engineering student at the University of Devon, in the last chapter. Her CV is shown on pages 154 and 155.

Angela's CV is mostly factual, yet it has been carefully thought out in order to emphasise her suitability for a particular job. The post that she is applying for is in general engineering and so she shows that she has some experience in practical engineering through her vacation jobs. It would not be wise to try to give any more prominence to these, since the periods were short. If she were applying for a job with a significant administrative component, she might give more information about her duties as Secretary of the University Harriers. Each time she uses her CV she will think about the appropriate emphasis. We talk about "Angela's CV" as if it were a definitive, unchanging document, but that should not be the case. The CV you submit for a job must demonstrate that you want to work for that particular organisation. "A good general CV" is not good enough. (So although we hope the example CVs in this chapter are helpful, you should not treat them as models.) It must be *your* document, conveying all that's best about you. If your CV catches the reader's interest, that will increase your chances of being interviewed and also to some extent set the tone of the interview. That will be good for you because you will be able to talk about things that interest you.

When you have an idea of overall strategy, think about the layout, and be fussy about everything. In presentation terms your CV must be as perfect as you can make it. A spelling mistake would be disastrous (and remember, computer spell-checks do not pick up all mistakes).

If something is given a lot of space it tends to appear more important to the reader. So Angela did not list her GCSEs in a vertical column, or give grades, as that would have given more prominence to them than to the later stages of her education.

Clearly the most relevant stage of your education is the most recent: your degree. You should refer to any particular achievements, honours or prizes. Your final year project or dissertation should also feature.

Most students have some work experience by the time they are applying for full-time jobs. Vacation work in your field, and better still the industrial training component of a sandwich degree, are obviously the most relevant, but it is also important to include other work experience, especially if it involves some responsibility. If you include a mixture of types of work experience, think carefully about the emphasis you give to each.

Angela's CV is set out in the usual way, in reverse chronological order. Employers expect this, and it gives more emphasis to recent achievements, which is usually a good tactic. In any case, be consistent: don't give your education in one order and your work experience in the other.

Jeff, who we met earlier in this chapter, has decided to draw attention to the period he spent running his own business before going to university. His CV is more descriptive than Angela's, not just letting the facts speak for themselves. Jeff's CV is on pages 156 and 157.

Angela and Jeff have chosen different formats in order to draw attention to different aspects of their lives and therefore make different statements to potential employers. Angela's CV says, "I've had a good traditional education, done some relevant vacation work, have varied interests, and am a good organiser". Jeff's says, "I've seen a bit of life, and I've already shown I can be successful in the real world".

There are other methods of setting out CVs. For example, a "functional CV" takes the emphasis right away from facts, and concentrates more on personal qualities. Your careers adviser will tell you more. The detailed layout on the page can also have many forms.

As we see from Angela's CV, leisure interests can help to show that you are an enterprising or interesting person. But if they don't show that, don't include them. (Jeff's hobby is riding motorbikes, and that wouldn't add much to the overall picture!)

You will need to name people who can write references for you. The normal number is two, but some employers may ask for three. One should be a lecturer on your course, preferably Personal Tutor or Course Leader. Ideally another should be your boss during a significant period of work experience.

Ask for permission first before you give someone's name. You want your referees to have a good opinion of you at the time they write the references, and an unexpected request from an unfamiliar source may threaten this state of mind. Obviously you should choose someone who you think will write you a good reference. (So it *does* matter what your lecturers think of you!)

Make sure you give details of your referees' names (with Dr, Professor, etc), job descriptions and addresses correctly. Give their phone numbers and email addresses if you can.

You should use a good printer and good-quality paper. Students' CVs tend to be two pages in length. Think carefully about where you break the text between the pages, and use all the space rather than stopping halfway down a sheet.

Take advantage of any opportunity to have your CV checked – by a careers adviser for example. Spend time responding to the comments. If you feel reluctant to show it to other people, it must need some more work. Most importantly, look upon these brief notes on CV writing as the beginning, not the end, of your education in CV preparation.

Letter of application

You may need to send a covering letter with your CV. One written by Angela is shown on page 153. She has highlighted some of the most relevant points in her CV, without being repetitive. In the letter, even more than the CV, you make the content specific: communicating your interest in working for the organisation and your suitability for the post. Any hint that the letter is "mass produced" will give a poor impression. Accuracy and appearance are very important; it is a good idea to use the same font for the letter as for the CV.

Other forms of application

Most large recruiters require applications to be made online. Typically a series of questions will require you to reflect on previous experience and give examples of skills. Think carefully about your answers. If you can, view all the questions before you complete the application. Give yourself time to compose and refine your answers. You may even be able to ask your careers adviser to check out your proposed answers. You can also practise these difficult skills on dedicated websites. When you complete an application online you may also be asked to upload your CV.

When completing a paper-based application form, many of the comments already made about CVs still apply. Opportunities for giving emphasis to particular aspects of your background still exist, but there is less scope.

There is a Standard Application Form (SAF) for graduate vacancies, which is accepted by some employers. Copies, and advice on completion, will be available from your careers service. Other companies require applications to be made on their own forms. If you have been asked to apply on an application form, don't send a CV instead, or fill in cursory details on the form and attach a CV.

Very often forms can be downloaded as a file. This means that it is easier to correct mistakes, but still needs thought and care. Since everyone's form will look similar, it's even more important that you think about how to make yours excellent.

As with a CV, be fussy. Take care over every detail, including simple things like the precise name of the company, of the vacancy, and of the institution at which you are studying.

Additional statement

The SAF gives the opportunity to expand on the basic facts in a number of places. Other application forms ask mainly for facts and then leave a space for "Further information in support of application" or something similar.

Remember that while it is important not to appear diffident or lacking in self-confidence, there is also a danger in coming across as arrogant and pushy, or simply too good to be true. Write down what you think your potential employer should know about you, without exaggeration or padding. The form may say, "continue on a separate sheet if necessary". Don't be afraid to do this, but don't add waffle just so you need to use the extra sheet. Whenever you can, give specific examples of achievements. Take care with the use of English. The additional statement is a major test of communication skills (and will be seen that way by employers too).

Applications for industrial placements and vacation work

The principles of applying for a sandwich training period, or vacation work, or any other industrial placement, are the same as those covered already in this chapter for other types of job. Since you have not completed your studies, there will probably be less emphasis on your degree and more on personal attributes than when you apply for a graduate position. These attributes are the same as for any job: commitment, energy, enthusiasm, common sense, initiative, resourcefulness, flexibility, and willingness to work hard, to work with others and to take responsibility. This may seem like a long list, but you don't need to have been the star performer every moment of your life to be offered a placement. If you are worth employing, you simply have to identify why. You will need to think carefully to find examples – in previous vacation or part-time work, or in your leisure interests – that demonstrate your attributes.

Applications for postgraduate study

Many graduates go on to further study or research. Of course in this case academic qualifications are particularly important. But personal attributes still matter: enthusiasm for the subject, general curiosity, and, particularly, capacity for independent study. Demonstrations of these should be easy to find in your academic experience so far. On your application form or CV you should give evidence of these qualities; for example, give information about your final year project or dissertation – not just the title, but a summary in a few sentences, or even a one-page abstract, including your conclusions and why they might be useful.

Checklist for a job application

- Have you made use of the careers advice facilities at your institution?
- Have you researched the organisation, thought about the sort of person they are looking for, and identified which of your skills are most relevant?
- Have you checked whether you should apply with a CV or an application form?
- Are your education details correct?
- Have you covered all relevant work experience?
- Have you shown yourself in the best light for this particular job?
- Have you been consistent, and fussy about details?
- Is your application tidy and smart?
- Is it well written?
- Have other people checked it?

Further reading

There are hundreds of books on applying for jobs. Pick one (from your library) that you like the look of, or preferably follow the advice from your careers service. They will guide you to valuable information on the Web, or may have guides and pamphlets (free or reasonably priced) specially for students.

18 Bere Crescent
Souweston
Devon SU8 2ER

11 May 2011

Mr G Parland
Personnel Manager
JJ Engineering
Enterprise Road
MANCHESTER M16 7TY

Dear Mr Parland

Vacancy for Graduate Engineer

I would like to be considered for the post of Graduate Engineer (reference E/563) as advertised in New Engineer magazine on 6 May. I attach my CV.

I will complete my BEng (Hons) Mechanical and Manufacturing Engineering degree at the University of Devon in July. In my final year I have been carrying out a project on the prediction of efficiency resulting from the introduction of manufacturing cells. I have organised a programme of visits to local manufacturing plants as part of my project studies. This has increased my interest in the practical aspects of manufacturing engineering. I also have practical engineering experience from two vacation jobs.

The declared goal of JJ Engineering to 'stay in the lead in the application of new technology' (from your website) appeals to me strongly. I have always enjoyed taking on challenges that combine technical innovation with practical implementation, and I have been able to demonstrate this on my course and in my work experience.

Yours sincerely

Angela Fulchrum

Angela Fulchrum

A.Fulchrum@devon.ac.uk

Angela Jane Fulchrum

ADDRESS (home)   54 Pampard Road
Croydon
Surrey CR2 4PR

EMAIL   A.Fulchrum@devon.ac.uk

TELEPHONE   0787 596 0339

EDUCATION AND QUALIFICATIONS

2008 to 2011   *University of Devon*

Course: **BEng (Hons), Mechanical and Manufacturing
Engineering**
Due to graduate July 2011.

Final year project: *Modelling of manufacturing efficiency.*
This has involved developing a novel method of modelling
manufacturing efficiency, and using it to predict the impact of
manufacturing cells. The outcomes are presented in a full report
and also on a summary sheet for local factory managers. At my
suggestion, visits to local factories were included in the project,
and I arranged these myself – five in total. All the manufacturing
managers I visited expressed an interest in receiving the
summary of my results.

2002 to 2008   *Haling High School, Croydon*

2008   A-levels: Maths (B), Physics (C), Geography (C)
2006   GCSE (7): Maths, Physics, Chemistry, English, French,
Geography, History

EMPLOYMENT

Summer Vacation   *Design Technician*
2010 (7 weeks)   *Dufin Engineering plc, Croydon*
I worked on small-scale alterations to hydraulic machinery for
new applications. I designed complex manifold systems, with
minimal supervision. I was encouraged to develop my own ideas
and I developed a number of new solutions. My final designs
were checked by an engineer when complete, but did not need
to be altered.

Summer Vacation **2009 (6 weeks)**	*Laboratory Technician (Electronic Engineering)* *University of Manitoba, Canada* This involved maintenance of equipment, and giving assistance to researchers undertaking practical work. I was given full responsibility for processing complex results from one two-week-long experiment.
INTERESTS	**Sport:** Athletics; Secretary, University Harriers; with responsibility for arranging and confirming fixtures, and organising transport, social events and the Annual General Meeting. **Music:** Singer in a folk group.
OTHER SKILLS	*Organisation* My role as Secretary of the University Harriers has developed my organisational skills considerably. The arrangements for events are complex, sometimes involving several other universities, and more than a hundred people. A great deal of tact is required, as students do not like to be organised. *Computing* Good familiarity with WordUp, Excite, Project+, Super-CAD/CAM, and programming in Dstar. Clean driving licence since May 2008.
REFEREES	Dr G H Bowen BEng Course Leader School of Engineering University of Devon Campus Park Souweston SU1 1AA 01111 453967 G.Bowen@devon.ac.uk

Mr I Saines
Senior Engineer
Dufin Engineering plc
Centre House
Albert Street
Croydon CR1 0HB
020 8549 3028
ian.saines@dufineng.co.uk

Jeffrey Almondi

ADDRESS
Flat 6
81 String Street
London
SE31 6YH

EMAIL
almondi86@hotshot.co.uk

TELEPHONE
0777 721 3886

SUMMARY
After completing my secondary education at college, I followed my first love at that time, and became a motorcycle courier. This gave me the opportunity to run my own business for two years, from which I gained a great deal of valuable experience. Seeking the variety and prospects of a professional career, I made the decision to enrol on the BSc (Hons) degree in Environmental Technology at the University of South London, and am now in my final year.

EDUCATION AND QUALIFICATIONS

2008 to 2011
University of South London
Studying for **BSc (Hons), Environmental Technology**
Due to graduate July 2011.
Final year project: *Sustainable materials use in motorcycles.*
This has considered strategies for more sustainable use of materials in the manufacture, use, maintenance and disposal of motorcycles. It has included a complete life cycle analysis, and multi-criterion analysis for assessment of relative sustainability of different materials. It concludes that with some changes to material use, motorcycles offer the best chance for personal transport to remain a viable option in the future.

2003 to 2005
2005
Orchard College, Crystal Palace
A-levels: Maths (B), Geography (C), Computing (C)

1999 to 2003
2003
Long Hall School, Lower Sydenham
GCSE (7): Maths, Physics, Computing, English, Italian,
              Geography, Art

EMPLOYMENT

July 06–Sept 08   *Almondi Couriers, Penge*

For two years before starting my university course, I ran a small motorcycle courier firm that I set up myself. I began working on my own, using contacts from previous periods of courier work. I slowly built up the business and, for the second year, employed between six and eight riders. My mother was also an employee, working from home taking orders and processing invoices.

I was an active rider throughout, but also worked on negotiating with clients, agreeing terms with riders, and managing accounts.

March 06–July 06   *Fasta Bikas, Sydenham*
Motorcycle courier

July 05–March 06   *Denton Deliveries, Forest Hill*
Motorcycle courier

July 04–Sept 04   *Alf's Bike Shop, Sydenham*
Shop assistant

SKILLS   *Enterprise and leadership*: I have set up a business from scratch, won clients, secured finance, employed other people, and taken full responsibility for the success of the venture. I have learnt a great deal about self-reliance, decision making, and delivering to clients (in every sense). The skills I have developed are applicable not only to running a motorcycle courier business, but to any challenge in which resourcefulness, drive, leadership and energy are requirements.

Business skills: I have acquired skills in the full range of activities involved in running a small business, including selling, negotiating, resolving disputes, book-keeping, and dealing with banks, insurance and tax.

Driving: Clean driving licence for cars and motorcycles since 2004.

REFEREES   Dr A B Hardness   Mr H Denton
BSc Course Leader   Director
School of Applied Science   Denton Deliveries
University of South London   Connection House
High Street   24 Acorn Road
London SE33 1FG   London SE31 8NN
020 8745 2323   020 7888 8899
A.Hardness@slondon.ac.uk   harry@dentondel.co.uk

18. Interviews

There are two parts to this chapter. The first continues the theme of job applications, by concentrating on the job interview. The second is firmly back in the world of college, considering interviews that are part of the assessment of project work.

18.1 Job interviews

Being invited for a job interview means that the first stage of your job application has been a success. You should feel pleased, but it is too early to celebrate; there's a major challenge ahead.

What is the nature of the challenge of an interview? It is not to outwit the interviewers; their aim is to find out what sort of person you are, and there is no point in trying to convince them that you are someone you are not. The main challenge is to ensure that by the end of the interview there is nothing good about you (and relevant to the job) that the interviewers are not aware of. This can sometimes be difficult because of nerves or shyness or loss of concentration.

As with spoken presentations, nerves are not necessarily a bad thing as long as they don't ruin your performance. Some advisers say "the interviewers will be as nervous as you are". This may be correct on occasion, but one thing is always true: the interviewers will all have been interviewed themselves for jobs in the past, will probably be interviewed in the future, and will know what it feels like.

The best way to counteract the damaging effects of nerves, and to improve your performance generally, is to *prepare well*.

Preparation

You will have carried out a great deal of the preparation at an earlier stage of your application (as described in the last chapter). You should possess information about the vacancy and about the organisation. Part of your preparation for the interview will involve looking back at the material you collected when you made your application, and looking again at the application itself.

When preparing for the interview, in the same way as when working on the application, you should try to make full use of the careers guidance facilities at your institution. Your careers advisers will give you tips on interview technique and suggest further reading. They will have videos on being interviewed which you should find helpful. They may even put you through a mock interview.

An alternative is to get a friend to give you a mock interview. You must choose this friend carefully; you must both be able to take the exercise seriously for it to be of value.

As well as this you can practise answering questions in your imagination. You can do this anywhere – walking down the street, sitting on the bus, lying in the bath. You should imagine yourself as the interviewer in order to think of obvious questions.

It is often said that the only really difficult questions are the ones you don't expect. Asking yourself testing questions (or being asked them by a friend) can help to prepare you for the unexpected. You might even imagine being asked some of the questions by an off-hand or aggressive interviewer, though this tactic is not particularly common. Most interviewers want to find out what you are really like, not make you more nervous. Also your interviewer might be your future boss and will not want to make an enemy of you at the start.

Here are some examples of questions. This is not a comprehensive list, it is just a starting point for you to make your own.

Interview questions

Why have you applied for this job?
What particularly interests you about it?
What makes you suitable for the job?
What are you seeking in any job?
What plans do you have for your career?
What excites you about a career in this field?
What talents/abilities will you bring to this organisation?
What are your strengths and weaknesses?
How would you describe your own personality?
What is special about you?
Give an example of your leadership abilities.
Tell us what you know about our organisation.
Who else are you applying to? Which job would you accept if you were offered them all?
What challenges will this organisation face in the next two years?

Many of the questions you will be asked will be based on your application. You should reread it thinking of likely questions. (Angela Fulchrum's covering letter, page 153, is likely to trigger something like "You say you have always enjoyed taking on challenges that combine technical innovation with practical implementation. Give some examples, and, if the outcome has been successful, tell us why.")

Asking yourself questions will give you practice in composing answers. You should not try to memorise an answer to any question, because it will sound unnatural when you repeat it. But phrases you have composed during one of your imaginary answers will come back to you during the interview and can be worked into what you say. You should be ready to be asked at the start of the interview to give a brief introduction about yourself.

When you prepare, you should bear in mind the qualities that you think the interviewers will be looking for. Remember that these will include professional qualities like education, experience and aptitude, and personal qualities like enthusiasm, positive attitude, and the ability to work with others. The interviewers will want to find someone who will do the job well, and someone who they will get on with.

The most effective form of interview practice is attending real interviews. This is certain to improve your technique, even if you might wish that you could get the perfect job first time. Try to be positive about rejection; you will always learn something. It is quite normal to contact the firm afterwards to ask for feedback on your interview performance.

The interview

The interview will be with one other person or with a panel. There may be two stages: a preliminary interview, followed (for those successful at the first stage) by a second interview.

In either case the programme may include "informal" sessions with all the interviewees together, at which you will be given information on the organisation or shown round. These are difficult times: you are obviously being assessed to some extent and being compared with the other applicants, but you should try to pace yourself, and not become worn out before the interview itself. The programme will probably have been organised in this way because it is more convenient for your hosts to organise certain activities for everyone together. There is no point in trying desperately to impress the people who show you round the building, when they may not be greatly involved in the selection in any case. Being shown round also allows you to think about whether you would be happy in the job. Remember that the whole thing is a two-way process.

Let's concentrate on the interview itself. You should arrive early, and should aim to eliminate any possibility of arriving flustered or disorganised. Report for the interview between five and ten minutes before time. If you get to the building much earlier, have a walk round the block, or sit in a park to compose your thoughts.

The first impressions that you make on the interviewers are important. You must be dressed smartly; you should have given this aspect particular attention, as students don't often need to be smart. The best assumption is that you cannot be too smart for an interview. As you walk into the room don't try to look artificially "dynamic", or let yourself look worried. Be cheerful and polite. If you have brought papers or samples of work, have everything well organised and ready to refer to.

Sit in an alert but comfortable position. Speak naturally: don't try to put on an artificial accent, but at the same time don't use colloquialisms or speak carelessly.

You will be asked at the end if you would like to ask any questions yourself. It is entirely appropriate for you to take up this opportunity, and you should make sure that you have found out everything important that you want to know. It is often wise not to ask about salary. (Your conditions of employment will be specified if you are offered the job.) You should make sure you fully understand the arrangements for training. Don't ask questions for the sake of it; if all your important queries have been answered during the day, say so.

Some students are a little daunted by the contrast between student life and the world of work, and fear that their interviewer may be rather scornful towards students. If this were true, the company would not be interviewing students or employing graduates. You certainly don't have to pretend to be old for your years or "boring", although an employer would be put off by immaturity.

Assessment centres

An assessment centre is a process not a place. In many ways it is an extension of the interview. Employers use assessment centres to see how an applicant reacts in realistic work situations. Typical activities include:

- group exercises (business case analysis, model bridge building, etc.)
- presentations
- problem solving
- psychometric tests (including mathematical, verbal and personality tests)
- role plays (sometimes using actors)
- further interviews.

The best advice is to be yourself and always try to contribute. Successful applicants are not necessarily those that push to be the "leader" in team exercises, but those who show honesty, empathy, supportive behaviour, good social and communication skills, and an aptitude for teamwork.

Checklist for preparing for a job interview

- Have you looked back over all the material you acquired when you first made your application (especially the application itself)?
- Have you thought again about the qualities the organisation will be looking for, and the qualities you have to offer?
- Have you made a list of likely (and unlikely) questions, and answered them yourself?
- Have you used friends and your careers service to help you prepare?
- Do you know when and where the interview will be held, how to get there, and how to ensure you will arrive on time?
- Are your smart clothes ready?

Interviews for industrial placements and vacation work

These may be more relaxed than interviews for full-time jobs, but still require serious preparation. If the interview has been set up via the college, you may not have needed to find out much about the company when you applied. But the interviewer will expect you to have done some research and will be disappointed if you haven't. "At least they could have had a look at our website", employers sometimes say.

The qualities that interviewers will be looking for have been discussed in Chapter 17. The obvious aspects of interview technique, for example smartness and punctuality, are just as important as they are for full-time interviews, because they show that you are serious about work. I once said to a student after his first placement interview that he would have a better chance if he wore a suit. He seemed surprised and said, "But they know I'm a student!" The point is that an employer will be looking particularly to see if you will be able to make the transition between student and employee. When people leave college and start work, they are ready for a big change in their lives. However, when students go on placement, there is a danger (in the employer's mind) that they may carry on considering themselves as students, reluctant to allow someone else to dictate their approach to work and timekeeping. Of course the employer will expect you to act like an employee, not a student; and, after all, that's the whole point of doing a placement – to experience real work.

Questions for placement interview
(in addition to those on page 159)

Why do you want a placement?
How do you think it would benefit you?
What sort of career do you want when you graduate?
What do you like about your degree studies?
What subjects are you good/bad at?
Do you know much about this company?
Since you don't have professional experience, how could you contribute to our work?
What do you think you would be doing if you had a placement with us?
Would you be able to use what you have learnt in college so far? What particularly?

18.2 Project interviews

Student projects are often partly assessed by interview or "viva" (short for *viva voce*, Latin for "with the living voice"). There may be important marks at stake, but in one sense these interviews are easier to cope with than job interviews. In a project interview the nature of the challenge is more easily defined: you are being assessed on fewer fronts; first impressions and selling yourself as a person are less significant.

The main purposes of project interviews are to determine:

that the work is yours
that you understand it
that you can explain it.

As with other "speaking tests" most people get nervous, but it should be possible to control nerves more in this type of interview than in a job interview or spoken presentation. The likely outcome of the interview is really determined long before the interview takes place. Either you have a good understanding of what you did in the project or you don't, and no amount of quick thinking or good performance in the interview is going to change that. So there is really nothing to be nervous about (unless you *don't* understand what you did).

Preparation

If a member of staff has given you personal supervision during your project, then that member of staff can help you prepare for the interview. To make full use of the opportunities, you will need to complete your report well before the interview. Then your supervisor should be able to comment on the strengths and weaknesses of your work, and perhaps give you an idea of the types of questions you are likely to be asked in the interview.

You can prepare yourself by reading the report again and writing down likely questions. You must have your own copy of the report. While you do this you should make sure that you understand, and could explain, every detail. (Don't just revise your favourite bits.) There is no need to prepare precise answers to specific questions since, as in job interviews, a pre-prepared answer will never sound natural. You should simply think about the sorts of comments you would make.

You are likely to be asked to give a summary of what you did and what you achieved. This request is very predictable, and yet it seems to take many students by surprise. Again it is not a good idea to learn a statement by heart, because the question could take many forms, for example: "How would you summarise your project to a non-expert?" or "What's the most important thing you found out?", or there could be an off-puttingly informal request like "Well, tell us what you've been up to".

The interview

A common format is to be interviewed by a small panel of staff, including some who have read your report and some who have not. Your supervisor will probably be there, as much to help you as to assess you. It would be worth asking your supervisor beforehand about who will be interviewing you.

Your interview will probably be fairly relaxed, but don't be put off by any apparent lack of friendliness. A few lecturers like to be stern or abrupt in these situations – they can't help it. Be cheerful yourself, and show your enthusiasm. Don't assume that a question asked in a seemingly aggressive way is a hard question, or one asked in a friendly way is an easy question. The interview is fundamentally about knowledge and understanding.

If you have a good grasp of the material, there is no reason why you should not look forward to your interview. You will be talking about something that interests you, and other people will be taking an interest in what you say. Your lecturers will have to listen to you for a change!

Further reading

See Chapter 17 (for job interviews).

19. What next?

This chapter has two purposes:

1. To tell you about some of the differences between communicating as a student and communicating as a professional.
2. To encourage you to continue improving your communication skills.

19.1 Professional communication

The main aim of this book has been to help you write and speak clearly while you are a student. Most of what you learn will be useful to you after you start work. However, you will then need to communicate in a number of additional ways, and some of the familiar ways may need a change of emphasis.

This section is intended as a brief introduction to these new areas. When you need to learn more you should refer to the suggestions in **Further reading**.

Spoken

Let's deal with communication in speech first. As a student, you must communicate clearly, and work well in a group, in certain fixed circumstances. But otherwise you have a great deal of freedom to work and to relate to others in the way that you choose. Also since most courses still place emphasis on written examinations, some of your most important challenges take place during three hours of complete silence!

In industry you will find a more constant requirement to communicate with others. You will be working directly with other people most of the time. Much of the communication will be informal, yet must allow the communication of technical information with precise clarity.

Even when you have found a job that you are happy with, you will not have seen the last of interviews. Many organisations have internal interviews for staff appraisal, salary negotiation and promotion. As you become more senior, you will find yourself on the other side of the interview table for internal interviews or for employing new staff. Interviewing effectively is a demanding task.

Your practice in making spoken presentations as a student will benefit you when you are in industry. You may be asked to make a presentation on a proposal to a client, or a public lecture or presentation to a conference. A specialised form of spoken presentation which may become your responsibility is the presentation of technical evidence to a court of law or a public inquiry.

Some of these forms of spoken presentation include the important component of communicating technical information to a non-technical audience. Modern society needs more technical people who are good at this.

Many business skills are forms of spoken communication. The art of negotiation is one of the most important.

Written

Not all written communication at work is formal. Notes, emails and faxes can be informal, and yet should always be clear and precise. The slightly more formal communication within an organisation is the **memo** (short for *memorandum*; plural, *memoranda*). This may be on company stationery, with a standard format at the top:

To:
From:
Date:
Subject:

You may need to spend time writing records. Meetings are recorded as detailed minutes (as described in Chapter 14) or in the form of summary reports. Records must be kept of progress on site, or safety procedures. Those involved with the day-to-day running of an operation or a project must keep diaries.

Many of the types of report written in industry (and by students) have already been mentioned in Chapter 9 (Reports). There are, however, some fundamental differences between the writing of reports by practitioners and by students.

1. When a student writes a report, the reader (the lecturer) usually knows at least as much about the subject as the writer. When professional people write reports they usually know more about the subject than the readers (which is partly why they write the report).
2. When reports are written in industry, the writer must take particular care to define the readership.
3. The readers of a professional report may not read the whole report. That is why it is important to write a good summary, and why it may be a good idea to place conclusions and recommendations near the start rather than the end. It is common to provide an executive summary (see page 84).
4. There may be standard formats for some types of report and standard points of style required by particular organisations.

19.2 Continuing to improve

Most people continue improving their communication skills throughout their working lives. This is the result of experiencing the effects of good and bad communication, and of thinking about personal strengths and weakness. The people who improve the most are the ones who really want to.

So . . . carry on learning for yourself, and aim to continue improving as a communicator. Don't underestimate the importance of this, or you may not achieve your full potential. Take responsibility for developing your own skills and think carefully about other people's advice. Collect your own good ideas, and keep up an interest in writing, speaking and language (see **Further reading**).

Professional people in virtually any job in industry will tell you, "That's what most of my job consists of: communication".

Further reading

On professional communication

Ellis, Richard *Constructive communication, skills for the building industry.* Arnold, 1999. Professional communication in construction.
van Emden, Joan *Writing for engineers*, 3rd edition. Palgrave Macmillan, 2005. Good advice on style and format for professional engineers.

On language and writing

Bryson, Bill *Mother Tongue.* Penguin, 1991. An entertaining and fascinating look at the history and usage of English.

The books recommended in Chapter 4.

Complete list of further reading suggestions

A dictionary

A thesaurus

A pocket-sized book on English usage

Higher Education Academy website sources on PDP

Back, Ken and Kate Back *Assertiveness at work: a practical guide to handling awkward situations*, 3rd edition. McGraw-Hill, 2005.

Beer, David and David McMurrey *A guide to writing as an engineer*, 3rd edition. Wiley, 2009.

Breach, Mark *Dissertation writing for engineers and scientists – student edition*. Prentice Hall, 2009.

Bryson, Bill *Mother Tongue*. Penguin, 1991.

Cottrell, *Stella skills for success – the personal development handbook*. Palgrave, 2003.

Edney, Andrew *PowerPoint 2007 in easy steps*. Computer Step, 2007.

Ellis, Richard *Constructive communication, skills for the building industry*. Arnold, 1999.

Kirkup, Les *Experimental methods: an introduction to the analysis and presentation of data*. Wiley, 1994.

Liengme, Bernard V. *Guide to Microsoft Excel 2007 for scientists and engineers*. Academic Press, 2009.

Naoum, S.G. *Dissertation research and writing for construction students*, 2nd edition. Butterworth-Heinemann, 2007.

Neville, Colin *The complete guide to referencing and avoiding plagiarism*, 2nd edition. Open University Press, 2010.

Northedge, Andrew *The good study guide*. Open University, 2005.

Seely, John *Oxford Guide to Effective Writing and Speaking*, 2nd edition. Oxford University Press, 2005.

Sides, Charles H. *How to write and present technical information*, 3rd edition. Cambridge University Press, 1999.

Lorraine Stefani, Robin Mason and Chris Pegler *The educational potential of e-portfolios – supporting personal development and reflective learning.* Routledge, 2007.

van Emden, Joan *Writing for engineers*, 3rd edition. Palgrave Macmillan, 2005.

Answers to tests

2a

All the answers appear in Chapter 2. Here's where to find them.

1. The correct spelling of all these words is given on page 10.
2. An observable fact; phenomena; Greek. Other tricky plurals are on pages 10 and 11.
3. Adverbs (and other parts of speech) are considered on page 9.
4. its/it's is discussed on page 11, the rest on pages 14 and 15.
5. See page 14.
6. See page 10.

2d

Other information to support application

My **principal** reason for applying for this post is that I feel that C G Freeland is a company that can be proud of **its** record in innovative design. Your new activities in **environmental** control (to which I **believe** I can contribute) perfectly **complement** your existing specialisms. I am interested in post E15 (Graduate), but would like to be considered for any **alternative** vacancies.

My main **leisure** activity is **swimming**. I have won county medals on no **fewer** than six **occasions**, and I represented England at the last Commonwealth Games.

3a

This is a piece of writing with no punctuation. What you must do is insert it. You may use commas, full stops, capital letters, paragraphs, or any other forms of punctuation that you think might be appropriate.

The main unit of written English is the sentence. Sentences can be long or they can be short. A sentence really expresses one thought. If the thought can be expressed in a brief statement, it is quite appropriate for the sentence to be short. However, some thoughts are more complex, and are linked together by words like "since", "and", "because" or "but". In technical English, where clarity is of prime importance, there is more danger in long sentences than short ones. Of course every sentence must contain a verb. If it doesn't, it is (in a manner of speaking) too short.

Paragraphs are also very useful for bringing clarity to written English. The break between paragraphs provides a definite pause in the text.

Punctuation really matters because it helps to make writing clear. People who do not write clearly, perhaps because their punctuation is poor, make a bad impression. They may be excellent at their profession in other respects, but if they fail to get good jobs, or fail to gain their clients' trust, they are likely to be disappointed in their careers.

3c

1. The number of members within a team depends on two factors: the size and complexity of the project.
2. A quality management system should be based on existing systems, amended and supplemented where necessary to conform with ISO 9001.
3. Control should be exercised throughout the whole process from start to finish; products within a subcontractor's work may have to be included.
4. What are the main problems with the current system?

3d

1. The electronics industry has been healthy compared with other industries; this can be clearly seen in the attached graphs.
2. This, coupled with high interest rates, has caused many small companies to fold.
3. Engineering will continue to be misunderstood, and we graduate engineers are the ones who will suffer most. (second comma removed)
4. The wall is relatively thin, but it is strengthened at regular intervals by buttress supports. (or remove comma)
5. Although these machines rarely need maintenance, do not have regular breaks like their human counterparts, and do not arrive late, they were not developed to replace humans.
6. Another interesting idea is one that is currently used in Houston, Texas. (first comma removed)
7. There is no requirement for the manager to be present. Isn't this unsatisfactory?

9a

Here is just one way of doing it.

1. Location
      (v) (t) (p) (f)
2. Hydrology
      (i) (n) (a) (q)
3. Environment and habitat
      (g) (j) (c)
4. Economic, social and political aspects
      (z) (x) (k) (r)

5. Ground conditions
    (o) (b) (u) (h)

6. Construction

    6.1 Materials
        (d)

    6.2 Access and facilities
        (s) (e) (l) (w)

    6.3 Environmental impact
        (m) (y)

9b

Here is a possible answer.

1. Introduction
(client, address of property, terms of reference)

2. Accommodation
    2.1 Summary (list of rooms that can be used)
    2.2 Room details (size, access, person-capacity, features)
    2.3 Other features (hall, stairs, balcony, garden . . .)

3. Facilities
    3.1 Kitchen (for food preparation, washing-up . . .)
    3.2 Sound equipment

4. Neighbours
    4.1 Within property (possibility for shared event, noise)
    4.2 Neighbouring property (noise limits)

5. Transport
    5.1 Parking
    5.2 Public transport
        5.2.1 Closeness to route(s)
        5.2.2 Timetable (frequency, last departure)
    5.3 Taxi fares

6. Recommendations

Index

For words in *italics*, the text gives advice on use of the actual word.